Creed Odyssey in Mathematics and Medicine series

Book 2 Three Open Problems by Riemann and Polignac

Book 2 Three Open Problems by Riemann and Polignac

Abstract

Creed Odyssey in Mathematics and Medicine series: *Book 2 Three Open Problems by Riemann and Polignac (Paperback ISBN: 9781710377835)*
Dedicated to my lovely daughter Jelena born 13 weeks early on May 14, 2012.
Cover design: Image of Great Pyramid of Giza designed with precision by ancient Egyptian Mathematicians.

CONTENTS	Page
Chapter 0 (PAPER) Mathematics for Incompletely Predictable Problems: Riemann zeta function and Sieve of Eratosthenes	2
Chapter 1 The Big Picture of Fundamental Laws	45
Chapter 2 Religion and Mitochondrion	49
Chapter 3 Beautiful Mathematics versus Sexy Mathematics	51
Chapter 4 Jelena 'Shinning Light'	54
Chapter 5 Medical Conspiracy and China Economic Miracle	56
About the Author: Professor John Ting	60

Mankind have always been fascinated by the beauty and power of mathematics. Ancient mathematicians from diverse parts of our planet such as China, India, Greece and the Arab world have acquired rich knowledge on different areas in mathematics.

Creed Odyssey in Mathematics and Medicine series authored by Dr. John Ting will reveal deep connections not just between Mathematics and Medicine but also between Mathematics, Physiology, Science and Religion. The ten chapters from **Book 1 Alphabet and Language of Science** outline all relevant materials emphasizing their crucial "Universal Language" role in building the innovative 'Spherical Model of Science' and 'Spherical Model of Numbers. The five chapters from **Book 2 Three Open Problems by Riemann and Polignac** introduce relevant materials for the three problems with a chapter dedicated to explaining Fundamental Laws. The six chapters from **Book 3 Rigorous Proofs for Three Open Problems** describe in layman's terms how rigorous proofs for the three problems are derived as per materials in the research paper on solving Riemann hypothesis, Polignac's and Twin prime conjectures [which are fully replicated at the beginning part Chapter 0 of Book 1, 2, 3 & 4]. The exotic A228186 Hybrid integer is also outlined in one of the chapters. The four chapters from **Book 4 Monstrous Moonshine and E-Commerce Apocalypse** contain materials that speculate on important role of Umbral ("Shadow"), Mathieu and Monstrous Moonshine in String theory potentially uniting Einstein's famous General Relativity and Quantum gravity, and refute a common misconception that solving Riemann hypothesis will lead to E-Commerce apocalypse.

In the research paper with title "*Mathematics for Incompletely Predictable Problems: Riemann zeta function and Sieve of Eratosthenes* published in viXra on Tuesday November 5, 2019": **Abstract** Mathematics for Incompletely Predictable Problems is associated with Incompletely Predictable problems containing Incompletely Predictable entities. Nontrivial zeros and two types of Gram points in Riemann zeta function together with prime and composite numbers from Sieve of Eratosthenes are Incompletely Predictable entities. Valid and complete mathematical arguments for first key step of converting this function into its continuous format version and second key step of applying Information-Complexity conservation to this Sieve are provided. Direct spin-offs from first step consist of proving Riemann hypothesis (and explaining both types of Gram points) and second step consist of proving Polignac's and Twin prime conjectures.

Creed Odyssey in Mathematics and Medicine series

1. Introduction

Mathematics for Pyramid.jpg

FIGURE 1. Mathematics is the abstract science of number, quantity and space, either as abstract concepts (pure mathematics), or as applied to other disciplines such as physics and engineering (applied mathematics). The ancient Egyptian mathematicians must possess significant knowledge in Algebra, Geometry and Calculus to successfully build the largely intact Great Pyramid of Giza which is the oldest of Seven Wonders of Ancient World.

Keywords:
Dirichlet Sigma-Power Laws, Exact and Inexact Dimensional analysis homogeneity, Information-Complexity conservation, Plus Gap 2 Composite Number Continuous Law, Plus-Minus Gap 2 Composite Number Alternating Law, Polignac's & Twin prime conjectures, Riemann hypothesis

Preliminary 1: This *Creed Odyssey in Mathematics and Medicine series* comprising of four well-organized books containing vast scientific knowledge is written in simple style catering for the general public as well as other experts such as in the fields of mathematics, medicine, science, physiology and religion. Some readers may elect to initially skip reading the Research paper at the beginning of each book. They may do so and return to read this Research paper later on without compromising their understanding of each book in its entirety.

Preliminary 2: A dedicated mathematician must have deep faith (creed) in his or her chosen mathematical field with intensity akin to how a dedicated priest must have deep faith (creed) in his or her chosen system of religious belief. A complete mathematical or religious journey (odyssey) usually consists of a long trip occurring over a large period of time involving many different and exciting activities. Bernhard Riemann (1826 – 1866) from Germany and Alphonse de Polignac (1817 – 1890) from France were considered dedicated mathematicians with creed odyssey in mathematics. Their mathematical contributions were immeasurable with Bernhard proposing Riemann hypothesis in 1859, and Alphonse proposing Twin prime conjecture in 1846 and Polignac's conjecture in 1849.

Preliminary 3: As a self-taught and dedicated mathematician, the author John Ting has literally traced the steps of mathematical discoveries taken by these two famous mathematicians when he successfully provided rigorous proofs for Riemann hypothesis, Polignac's and Twin prime conjectures in 2019. Possessing creed odyssey in mathematics and medicine, this monumental achievement is possible only after taking a significant time interval of approximately three years from 2016 to 2019 to do so. As a bonus, solving Riemann hypothesis has also resulted in explaining the closely related two types of Gram points.

Known from time immemorial, two mathematical quantities are in golden ratio if their ratio is the same as ratio of their sum to the larger of the two quantities. Mathematicians since Euclid have studied properties of golden ratio, including its appearance in dimensions of a regular pentagon and in a golden rectangle, which may be cut into a square and a smaller rectangle with the same aspect ratio. The golden ratio has also been used to analyze proportions of natural objects as well as man-made systems such as financial markets. The golden ratio appears in some patterns in nature, including the spiral arrangement of leaves and other plant parts.

In this research paper, treatise on relevant Mathematics for Incompletely Predictable Problems required to solve Riemann hypothesis and explain the closely related Gram points is outlined first; and to solve Polignac's and Twin prime conjectures is outlined subsequently.

Elements of three complete sets constituted by nontrivial zeros and two types of Gram points together with elements of two complete sets constituted by prime and composite numbers are all defined as Incompletely Predictable entities. Riemann hypothesis (1859) proposed all nontrivial zeros in Riemann zeta function to be located on its critical line. Defined as Incompletely Predictable problem is essential in obtaining the continuous format version of [discrete format] Riemann zeta function dubbed Dirichlet Sigma-Power Law to prove this hypothesis. Both types of Gram points when geometrically depicted as their corresponding x- and y-axes intercepts help explain that they will explicitly form integral parts of this function. Defined as Incompletely Predictable problems is essential for these explanations to be correct. Involving proposals that prime gaps and associated sets of prime numbers are infinite in magnitude, Twin prime conjecture (1846) deals with even prime gap 2 thus forming a subset of Polignac's conjecture (1849) which deals with all even prime gaps 2, 4, 6, 8, 10,.... Defined as Incompletely Predictable problems is essential to prove these conjectures using research method dubbed Information-Complexity conservation. Then Mathematics for Incompletely Predictable Problems equates to *sine qua non* defining problems involving Incompletely Predictable entities to be Incompletely Predictable problems achieved by incorporating certain identifiable mathematical steps with this procedure ultimately enabling us to rigorously prove or explain all the above mentioned problems.

Refined information on Incompletely Predictable entities of Gram and virtual Gram points: These entities of infinite magnitude are dependently calculated using *complex equation* Riemann zeta function, $\zeta(s)$, or its *proxy* Dirichlet eta function, $\eta(s)$, in critical strip (denoted by $0 < \sigma < 1$) thus forming integral parts of this function. In Figure 3 below, Gram[y=0], Gram[x=0] and Gram[x=0,y=0] points are respectively geometrical x-axis, y-axis and Origin intercepts at critical line (denoted by $\sigma = \frac{1}{2}$). Gram[y=0] and Gram[x=0,y=0] points are respectively synonymous with traditional 'Gram points' and *nontrivial zeros*. In Figures 4 and 5 below, virtual Gram[y=0] and virtual Gram[x=0] points are respectively geometrical x-axis and y-axis intercepts at non-critical lines (denoted by $\sigma \neq \frac{1}{2}$). Virtual Gram[x=0,y=0] points do not exist.

Refined information on Incompletely Predictable entities of prime and composite numbers: These entities of infinite magnitude are dependently computed (respectively) directly and indirectly using *complex algorithm* Sieve of Eratothenes. Denote $\mathbb{C}$ to be uncountable complex numbers, **R** to be uncountable real numbers, **Q** to be countable rational numbers or roots [of non-zero polynomials], **R–Q** to be uncountable irrational numbers, **A** to be countable algebraic numbers, **R–A** to be uncountable transcendental numbers, **Z** to be countable integers, **W** to be countable whole numbers, **N** to be countable natural numbers, **E** to be countable even numbers, **O** to be countable odd numbers, **P** to be countable prime numbers, and **C** to be countable composite numbers. **A** are $\mathbb{C}$ (including **R**) that are countable rational or irrational roots. We have (i) Set **N** = Set **E** + Set **O**, (ii) Set **N** = Set **P** + Set **C** + Number '1', (iii) Set **A** = Set **Q** + Set **irrational roots**, and (iv) Set **N** $\subset$ Set **W** $\subset$ Set **Z** $\subset$ Set **Q** $\subset$ Set **R** $\subset$ Set $\mathbb{C}$.

Then Set **R**–**Q** = Set **irrational roots** + Set **R**–**A**. With increasing size, arbitrary Set **X** can be countable finite set (CFS), countable infinite set (CIS) or uncountable infinite set (UIS). Cardinality of Set **X**, |**X**|, measures "number of elements" in Set **X**. E.g. Set **negative Gram[y=0] point** has CFS of negative Gram[y=0] point with |**negative Gram[y=0] point**| = 1, Set **even P** has CFS of even **P** with |**even P**| = 1, Set **N** has CIS of **N** with |**N**| = $\aleph_0$, and Set **R** has UIS of **R** with |**R**| = $\mathfrak{c}$ (cardinality of the continuum).

Differentiation of terms "Incompletely Predictable" versus "Completely Predictable": Set **N** = Set **E** + Set **O**. The two subsets of even and odd numbers are "Independent" and "Completely Predictable". Examples: The even number after 2,984 viz. 2,984 / 2 = 1,492nd even number is [easily] calculated independently using *simple algorithm* to be 2,984+2 = 2,986 viz. 2,986 / 2 = 1,493rd even number. The odd number after 2,985 viz. (2,985+1) / 2 = 1,493rd odd number is [easily] calculated independently using *simple algorithm* to be 2,985+2 = 2,987 viz. (2,987+1) / 2 = 1,494th odd number. Set **N** = Set **P** + Set **C** + Number '1'. The two subsets of prime and composite numbers are "Dependent" and "Incompletely Predictable". Example: The sixth prime number '13' [after fifth prime number '11'] is [not easily] computed dependently using *complex algorithm* from scratch via: 2 is 1st prime number, 3 is 2nd prime number, 4 is 1st composite number, 5 is 3rd prime number, 6 is 2nd composite number, 7 is 4th prime number, 8 is 3rd composite number, 9 is 4th composite number, 10 is 5th composite number, 11 is 5th prime number, 12 is 6th composite number, and our desired 13 is 6th prime number.

Formal definitions for Completely Predictable (CP) entities and Incompletely Predictable (IP) entities: In this paper, the word "number" [singular noun] or "numbers" [plural noun] in reference to prime & composite numbers, nontrivial zeros & two types of Gram points can interchangeably be replaced with the word "entity" [singular noun] or "entities" [plural noun]. Respectively, an IP (CP) number is locationally defined as a number whose position is *dependently* (*independently*) determined by complex (simple) calculations using complex (simple) equation or algorithm with (without) needing to know related positions of all preceding numbers in neighborhood. Simple properties are inferred from a sentence such as "This simple equation or algorithm by itself will intrinsically incorporate actual location [and actual positions] of all CP numbers". Solving CP problems with simple properties amendable to *simple* treatments using *usual* mathematical tools such as Calculus result in 'Simple Elementary Fundamental Laws'-based solutions. Complex properties, or "meta-properties", are inferred from a sentence such as "This complex equation or algorithm by itself will intrinsically incorporate actual location [but not actual positions] of all IP numbers". Solving IP problems with complex properties amendable to *complex* treatments using *unusual* mathematical tools such as Information-Complexity conservation, exact and inexact Dimensional analysis homogeneity as well as using *usual* mathematical tools such as Calculus result in 'Complex Elementary Fundamental Laws'-based solutions.

Based on Mathematics for Incompletely Predictable Problems, we compare and contrast CP entities (obeying Simple Elementary Fundamental Laws) against IP entities (obeying Complex Elementary Fundamental Laws) using examples:
(I) **E** are CP entities constituted by CIS of **Q** 2, 4, 6, 8, 10, 12....
(II) **O** are CP entities constituted by CIS of **Q** 1, 3, 5, 7, 9, 11....
(III) **P** are IP entities constituted by CIS of **Q** 2, 3, 5, 7, 11, 13....
(IV) **C** are IP entities constituted by CIS of **Q** 4, 6, 8, 9, 10, 12....

(V) With values traditionally given by parameter t, nontrivial zeros in Riemann zeta function are IP entities constituted by CIS of **R–A** [rounded off to six decimal places]: 14.134725, 21.022040, 25.010858, 30.424876, 32.935062, 37.586178,....

(VI) Traditional 'Gram points' (or Gram[y=0] points) are x-axis intercepts with choice of index 'n' for 'Gram points' historically chosen such that first 'Gram point' [by convention at n = 0] corresponds to the t value which is larger than (first) nontrivial zero located at t = 14.134725. 'Gram points' are IP entities constituted by CIS of **R–A** [rounded off to six decimal places] with the first six given at n = -3, t = 0; at n = -2, t = 3.436218; at n = -1, t = 9.666908; at n = 0, t = 17.845599; at n = 1, t = 23.170282; at n = 2, t = 27.670182.

Denoted by parameter t; nontrivial zeros, 'Gram points' and Gram[x=0] points all belong to well-defined CIS of **R–A** which will twice obey the relevant location definition [in CIS of **R–A** themselves and in CIS of numerical digits after decimal point of each **R–A**]. First and only negative 'Gram point' (at n = -3) is obtained by substituting CP t = 0 resulting in $\zeta(\frac{1}{2} + it) = \zeta(\frac{1}{2}) = -1.4603545$, a **R–A** number [rounded off to seven decimal places] calculated as a limit similar to limit for Euler-Mascheroni constant or Euler gamma with its precise (1st) position only determined by computing positions of all preceding (nil) 'Gram point' in this case. '0' and '1' are special numbers being neither **P** nor **C** as they represent nothingness (zero) and wholeness (one). In this setting, the ideas of (i) having factors for '0' and '1', or (ii) treating '0' and '1' as CP or IP numbers, is meaningless. All entities derived from well-defined simple/complex algorithms or equations are "dual numbers" as they can be simultaneously depicted as CP and IP numbers. For instance, **Q** '2' as **P** (& **E**), '97' as **P** (& **O**), '98' as **C** (& **E**), '99' as **C** (& **O**); CP '0' values in x=0, y=0 & simultaneous x=0,y=0 associated with various IP t values in $\zeta(s)$.

1.1 Algebraic versus Analytic number theory

Set **P** ⊂ Set **Z** ⊂ Set **Q**. Gaussian rationals, and Gaussian integers are complex numbers whose real and imaginary parts are (respectively) both rational numbers, and integer numbers. Gaussian primes are Gaussian integers z = a + bi satisfying one of the following properties.

1. If both a and b are nonzero then, a+bi is a Gaussian prime iff $a^2 + b^2$ is an ordinary prime.
2. If a = 0, then bi is a Gaussian prime iff |b| is an ordinary prime and |b| = 3 (mod 4).
3. If b = 0, then a is a Gaussian prime iff |a| is an ordinary prime and |a| = 3 (mod 4).

Prime numbers which are also Gaussian primes are 3, 7, 11, 19, 23, 31, 43,.... In Eq. (1) below, we noted that the equivalent Euler product formula with product over prime numbers [instead of summation over natural numbers] faithfully represent Riemann zeta function, $\zeta(s)$. Eq. (2) below is Riemann's functional equation involving transcendental number π (= 3.14159...). With denominators on the left involving odd numbers and named after Gottfried Leibniz, Leibniz formula for π states that $\frac{1}{1} - \frac{1}{3} + \frac{1}{5} - \frac{1}{7} + \frac{1}{9} - \cdots = \frac{\pi}{4}$. Expressions such as $\zeta(2) = \frac{1}{1^2} + \frac{1}{2^2} + \frac{1}{3^2} + \cdots = \frac{\pi^2}{6} \approx 1.64493406684822643647$ also involves π.

Algebraic number theory is loosely defined to deal with new number systems involving Completely Predictable or Incompletely Predictable entities such as even & odd numbers, prime & composite numbers, p-adic numbers, Gaussian primes, Gaussian rationals & integers, and complex numbers. A p-adic number is an extension of the field of rationals such that congruences modulo powers of a fixed prime number p are related to proximity in so-called "p-adic metric". The extension is achieved by an alternative interpretation of concept of "closeness" or absolute value viz. p-adic numbers are considered to be close when their difference is divisible by a high power of p: the higher the power, the closer they are. This property enables p-adic numbers to encode congruence information in a way that turns out to have powerful applications in number theory including, for example, attacking certain Diophantine equations and in the famous proof of Fermat's Last Theorem by English mathematician Sir Andrew John Wiles in 1995.

Analytic number theory is loosely defined to deal with functions of a complex variable such as Riemann zeta function [containing nontrivial zeros and two types of Gram points] and other L-functions. Study of prime numbers, complex numbers and π being braided together in a pleasing trio is usefully visualized to be located at the intersection of these two main branches of number theory. We separate our relatively elementary proof for Riemann hypothesis and relatively elementary explanations for two types of Gram points to belong to Analytic number theory, and our relatively elementary proofs for Polignac's and Twin prime conjectures [expectedly associated with paucity of functions of a complex variable] to belong to Algebraic number theory.

Indirect spin-offs from solving Riemann hypothesis are often stated as "With this one solution, we have proven five hundred theorems or more at once". This apply to many important theorems in Number theory (mostly on prime numbers) that rely on properties of Riemann zeta function such as where trivial and nontrivial zeros are / are not located. A classical example is resulting absolute and full delineation of prime number theorem, which relates to prime counting function. This function, usually denoted by $\pi(x)$, is defined as the number of prime numbers $\leqslant$ x. Public-key cryptography that is widely required for financial security in E-Commerce traditionally depend on solving the difficult problem of factoring prime numbers for astronomically large numbers. The intrinsic "Incompletely Predictable" property present in prime numbers, composite numbers, nontrivial zeros and two types of Gram points can never be altered to "Completely Predictable" property. For this stated reason, it is a mathematical impossibility that providing rigorous proofs such as for Riemann hypothesis will ever result in crypto-apocalypse. However, fast supercomputers and the far-more-powerful quantum computers that theoretically allow solving difficult factorization problem in quick time will result in less secure encryption and decryption. Then using, for instance, quantum cryptography that relies on principles of quantum mechanics to encrypt data and transmit it in a way that cannot be hacked will combat this issue.

Proposed by German mathematician Bernhard Riemann (September 17, 1826 – July 20, 1866) in 1859, Riemann hypothesis is mathematical statement on $\zeta(s)$ that critical line denoted by $\sigma = \frac{1}{2}$ contains complete Set **nontrivial zeros** with |**nontrivial zeros**| = $\aleph_0$. Alternatively, this hypothesis is geometrical statement on $\zeta(s)$ that generated curves when $\sigma = \frac{1}{2}$ contain complete Set **Origin intercepts** with |**Origin intercepts**| = $\aleph_0$.

$$\zeta(s) = \frac{e^{(\ln(2\pi)-1-\frac{\gamma}{2})s}}{2(s-1)\Gamma(1+\frac{s}{2})} \Pi_\rho \left(1 - \frac{s}{\rho}\right) e^{\frac{s}{\rho}} = \pi^{\frac{s}{2}} \frac{\Pi_\rho \left(1 - \frac{s}{\rho}\right)}{2(s-1)\Gamma\left(1+\frac{s}{2}\right)}$$

Depicted in full and abbreviated version, Hadamard product above is infinite product expansion of $\zeta(s)$ based on Weierstrass's factorization theorem displaying a simple pole at s = 1. It contains both trivial & nontrivial zeros indicating their common origin from $\zeta(s)$. Set **trivial zeros** occurs at σ = -2, -4, -6, -8, -10,..., ∞ with |**trivial zeros**| = $\aleph_0$ due to Γ function term in denominator. Nontrivial zeros occur at $s = \rho$ with γ denoting Euler-Mascheroni constant.

Remark 1.1. Confirming first 10,000,000,000,000 nontrivial zeros location on critical line implies but does not prove Riemann hypothesis to be true.

Locations of first 10,000,000,000,000 nontrivial zeros on critical line have previously been computed to be correct. Hardy in 1914[1], and with Littlewood in 1921[2], showed infinite nontrivial zeros on critical line by considering moments of certain functions related to $\zeta(s)$. This discovery cannot constitute rigorous proof for Riemann hypothesis because they have not exclude theoretical existence of nontrivial zeros located away from this line.

1.2 Exact & inexact Dimensional analysis homogeneity for equations & inequations

Respectively for 'base quantities' such as *length*, *mass* and *time*; their fundamental SI 'units of measurement' meter (m) is defined as distance travelled by light in vacuum for time interval $1/299\ 792\ 458$ s with speed of light c = 299,792,458 ms^{-1}, kilogram (kg) is defined by taking fixed numerical value Planck constant h to be 6.626 070 15 X 10^{-34} Joules·second (Js) [whereby Js is equal to kgm^{2}s^{-1}] and second (s) is defined in terms of ΔvCs = $\Delta(^{133}$Cs$)_{hfs}$ = 9,192,631,770 s^{-1}. Derived SI units such as J and ms^{-1} respectively represent 'base quantities' *energy* and *velocity*. The word 'dimension' is commonly used to indicate all those mentioned 'units of measurement' in well-defined equations.

Dimensional analysis (DA) is an analytic tool with DA homogeneity and non-homogeneity (respectively) denoting valid and invalid equation occurring when 'units of measurements' for 'base quantities' are "balanced" and "unbalanced" across both sides of the equation. E.g. equation 2 m + 3 m = 5 m is valid and equation 2 m + 3 kg = 5 mkg is invalid (respectively) manifesting DA homogeneity and non-homogeneity.

Remark 1.2. We can validly apply useful concepts from exact and inexact Dimensional analysis homogeneity to well-defined equations and inequations.

Let (2n) and (2n-1) be 'base quantities' in our derived Dirichlet Sigma-Power Laws formatted in simplest forms as equations and inequations. E.g. DA on exponent $\frac{1}{2}$ in $(2n)^{\frac{1}{2}}$ in simplest form is correct but DA on exponent $\frac{1}{4}$ in equivalent $(2^2 n^2)^{\frac{1}{4}}$ *not* in simplest form is incorrect. Fractional exponents as 'units of measurement' given by $(1-\sigma)$ for equations and $(\sigma+1)$ for inequations when $\sigma = \frac{1}{2}$ coincide with exact DA homogeneity[1]; and $(1-\sigma)$ for equations and $(\sigma+1)$ for inequations when $\sigma \neq \frac{1}{2}$ coincide with inexact DA homogeneity[2]. Respectively for equations and inequations, exact DA homogeneity at $\sigma = \frac{1}{2}$ denotes $\sum$(all fractional exponents) as $2(1-\sigma)$ and $2(\sigma+1)$ equates to ["exact"] whole number '1' and '3'; and inexact DA homogeneity at $\sigma \neq \frac{1}{2}$ denotes $\sum$(all fractional exponents) as $2(1-\sigma)$ and $2(\sigma+1)$ equates to ["inexact"] fractional number '$\neq 1$' and '$\neq 3$'.

Footnote 1,2: Exact and inexact DA homogeneity occur in Dirichlet Sigma-Power Laws as equations or inequations for Gram[y=0] points, Gram[x=0] points and nontrivial zeros. *Law of Continuity* is a heuristic principle *whatever succeed for the finite, also succeed for the infinite*. Then these Laws which inherently manifest themselves on finite and infinite time scale should "succeed for the finite, also succeed for the infinite".

Outline of proof for Riemann hypothesis. To simultaneously satisfy two mutually inclusive conditions: I. *With rigid manifestation of exact DA homogeneity*, Set **nontrivial zeros** with |**nontrivial zeros**| = $\aleph_0$ is located on critical line (viz. $\sigma = \frac{1}{2}$) when $2(1-\sigma)$ [or $2(\sigma+1)$] as $\sum$(all fractional exponents) = whole number '1' [or '3'] in Dirichlet Sigma-Power Law[3] as equation [or inequation]. II. *With rigid manifestation of inexact DA homogeneity*, Set **nontrivial zeros** with |**nontrivial zeros**| = $\aleph_0$ is not located on non-critical lines (viz. $\sigma \neq \frac{1}{2}$) when $2(1-\sigma)$ [or $2(\sigma+1)$] as $\sum$(all fractional exponents) = fractional number '$\neq 1$' [or '$\neq 3$'] in Dirichlet Sigma-Power Law[3] as equation [or inequation].

Footnote 3: Derived from original $\eta(s)$ (*proxy* for $\zeta(s)$) as equation or inequation, this Law symbolizes end-result proof on Riemann hypothesis.

Riemann hypothesis mathematical foot-prints. Six identifiable steps to prove Riemann hypothesis: *Step 1* Use $\eta(s)$, *proxy* for $\zeta(s)$, in critical strip. *Step 2* Apply Euler formula to $\eta(s)$. *Step 3* Obtain "simplified" Dirichlet eta function which intrinsically incorporates *actual location [but not actual positions]* of all nontrivial zeros[4]. *Step 4* Apply Riemann integral to "simplified" Dirichlet eta function in discrete (summation) format. *Step 5* Obtain Dirichlet Sigma-Power Law in continuous (integral) format as equation or inequation. *Step 6* Note exact and inexact DA homogeneity on their fractional exponents.

Footnote 4: Respectively Gram[y=0] points, Gram[x=0] points and nontrivial zeros are Incompletely Predictable entities with actual positions determined by setting $\sum Im\{\eta(s)\} = 0$, $\sum Re\{\eta(s)\} = 0$ and $\sum ReIm\{\eta(s)\} = 0$ to *dependently* calculate relevant positions of all preceding entities in neighborhood. Respectively actual location of Gram[y=0] points, Gram[x=0] points and nontrivial zeros; and virtual Gram[y=0] points, virtual Gram[x=0] points and "absent" nontrivial zeros occur precisely at $\sigma = \frac{1}{2}$; and $\sigma \neq \frac{1}{2}$.

2. Riemann zeta and Dirichlet eta functions

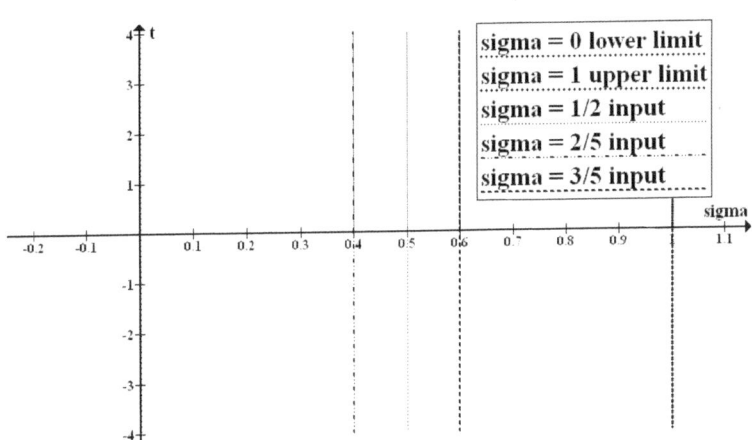

FIGURE 2. INPUT for $\sigma = \frac{1}{2}, \frac{2}{5}$, and $\frac{3}{5}$. $\zeta(s)$ has countable infinite set of Completely Predictable trivial zeros at $\sigma = $ all negative even numbers and countable infinite set of Incompletely Predictable nontrivial zeros at $\sigma = \frac{1}{2}$ for various t values.

L-functions form an integral part of 'L-functions and Modular Forms Database' (LMFDB) with far-reaching implications. In perspective, $\zeta(s)$ is simplest example of an L-function. $\zeta(s)$ is a function of complex variable s $(= \sigma \pm \imath t)$ that analytically continues sum of infinite series $\zeta(s) = \sum_{n=1}^{\infty} \frac{1}{n^s} = \frac{1}{1^s} + \frac{1}{2^s} + \frac{1}{3^s} + \cdots$. The common convention is to write s as $\sigma + \imath t$ with $\imath = \sqrt{-1}$, and σ and t real. Valid for $\sigma > 0$, we write $\zeta(s)$ as $Re\{\zeta(s)\} + \imath \cdot Im\{\zeta(s)\}$ and note that $\zeta(\sigma + \imath t)$ when $0 < t < +\infty$ is the complex conjugate of $\zeta(\sigma - \imath t)$ when $-\infty < t < 0$.

Also known as alternating zeta function, $\eta(s)$ must act as *proxy* for $\zeta(s)$ in critical strip (viz. $0 < \sigma < 1$) containing critical line (viz. $\sigma = \frac{1}{2}$) because $\zeta(s)$ only converges when $\sigma > 1$. This implies $\zeta(s)$ is undefined to left of this region in critical strip which then requires $\eta(s)$ repre-

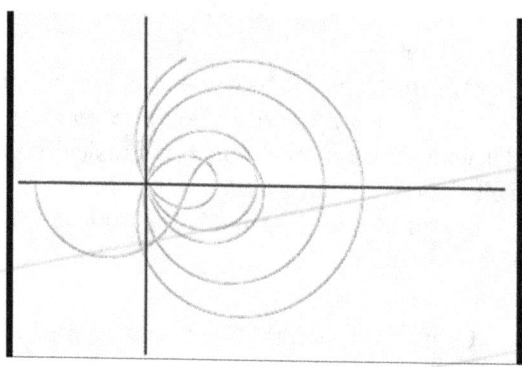

FIGURE 3. OUTPUT for $\sigma = \frac{1}{2}$. Schematically depicted polar graph of $\zeta(\frac{1}{2}+it)$ plotted along critical line for real values of t running from 0 to 34, horizontal axis: $Re\{\zeta(\frac{1}{2}+it)\}$, and vertical axis: $Im\{\zeta(\frac{1}{2}+it)\}$. There are presence of Origin intercepts which are totally absent in Figures 4 and 5 [with identical axes definitions].

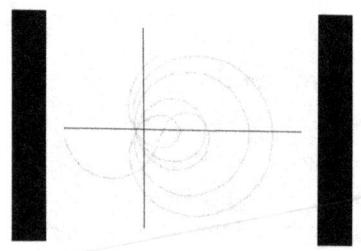
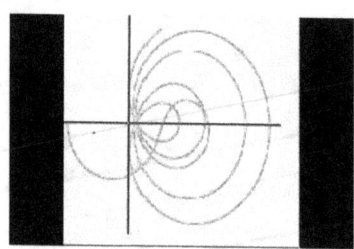

FIGURE 4. OUTPUT for $\sigma = \frac{2}{5}$. FIGURE 5. OUTPUT for $\sigma = \frac{3}{5}$.

sentation instead. They are related to each other as $\zeta(s) = \gamma \cdot \eta(s)$ with proportionality factor $\gamma = \frac{1}{(1-2^{1-s})}$ and $\eta(s) = \sum_{n=1}^{\infty} \frac{(-1)^{n+1}}{n^s} = \frac{1}{1^s} - \frac{1}{2^s} + \frac{1}{3^s} - \cdots$.

$$\zeta(s) = \sum_{n=1}^{\infty} \frac{1}{n^s} \qquad (1)$$
$$= \frac{1}{1^s} + \frac{1}{2^s} + \frac{1}{3^s} + \cdots$$
$$= \Pi_{p\ prime} \frac{1}{(1-p^{-s})}$$
$$= \frac{1}{(1-2^{-s})} \cdot \frac{1}{(1-3^{-s})} \cdot \frac{1}{(1-5^{-s})} \cdot \frac{1}{(1-7^{-s})} \cdot \frac{1}{(1-11^{-s})} \cdots \frac{1}{(1-p^{-s})} \cdots$$

Eq. (1) is defined for only $1 < \sigma < \infty$ region where $\zeta(s)$ is absolutely convergent. There are no zeros located here. In Eq. (1), equivalent Euler product formula with product over prime numbers [instead of summation over natural numbers] can also represent $\zeta(s)$.

$$\zeta(s) = 2^s \pi^{s-1} \sin\left(\frac{\pi s}{2}\right) \cdot \Gamma(1-s) \cdot \zeta(1-s) \qquad (2)$$

With $\sigma = \frac{1}{2}$ as symmetry line of reflection, Eq. (2) is Riemann's functional equation valid for $-\infty < \sigma < \infty$. It can be used to find all trivial zeros on horizontal line at $\imath t = 0$ occurring when $\sigma =$ -2, -4, -6, -8, -10,..., ∞ whereby $\zeta(s) = 0$ because factor $\sin(\frac{\pi s}{2})$ vanishes. Γ is gamma function, an extension of factorial function [a product function denoted by ! notation whereby n! $= n(n-1)(n-2)\ldots(n-(n-1))$] with its argument shifted down by 1, to real and complex numbers. That is, if n is a positive integer, $\Gamma(n) = (n-1)!$

$$\zeta(s) = \frac{1}{(1-2^{1-s})} \sum_{n=1}^{\infty} \frac{(-1)^{n+1}}{n^s} \qquad (3)$$

$$= \frac{1}{(1-2^{1-s})} \left(\frac{1}{1^s} - \frac{1}{2^s} + \frac{1}{3^s} - \cdots \right)$$

Eq. (3) is defined for all $\sigma > 0$ values except for simple pole at $\sigma = 1$. As alluded to above, $\zeta(s)$ without $\frac{1}{(1-2^{1-s})}$ viz. $\sum_{n=1}^{\infty} \frac{(-1)^{n+1}}{n^s}$ is $\eta(s)$. It is a holomorphic function of s defined by analytic continuation and is mathematically defined at $\sigma = 1$ whereby analogous trivial zeros with presence only for $\eta(s)$ [but not for $\zeta(s)$] on vertical straight line $\sigma = 1$ are found at $s = 1 \pm \imath \cdot \frac{2\pi k}{\ln(2)}$ where k = 1, 2, 3, 4, 5, ..., ∞.

Figure 2 above depict complex variable s ($= \sigma \pm \imath t$) as INPUT with x-axis denoting real part Re{s} equating to σ, and y-axis denoting imaginary part Im{s} equating to t. Figures 3, 4 and 5 above respectively depict $\zeta(s)$ as OUTPUT for real values of t running from 0 to 34 at $\sigma = \frac{1}{2}$ (critical line), $\sigma = \frac{2}{5}$ (non-critical line), and $\sigma = \frac{3}{5}$ (non-critical line) with x-axis denoting real part Re$\{\zeta(s)\}$ and y-axis denoting imaginary part Im$\{\zeta(s)\}$. There are infinite types-of-spirals possibilities associated with each σ value arising from all infinite σ values in critical strip. Mathematically proving all nontrivial zeros location on critical line as denoted by solitary $\sigma = \frac{1}{2}$ value equates to geometrically proving all Origin intercepts occurrence at solitary $\sigma = \frac{1}{2}$ value. Both result in rigorous proof for Riemann hypothesis.

3. Prerequisite lemma, corollary and propositions for Riemann hypothesis

Original equation $\eta(s)$, *proxy* for $\zeta(s)$, is treated as unique mathematical object with key properties and behaviors. Containing all x-axis, y-axis and Origin intercepts, it will intrinsically incorporate *actual location [but not actual positions]* of all Gram[y=0] points, Gram[x=0] points and nontrivial zeros. Proofs on lemma, corollary and propositions on nontrivial zeros depict exact and inexact DA homogeneity in both derived equation and inequation. Parallel procedure on Gram[y=0] and Gram[x=0] points below depict exact and inexact DA homogeneity in similarly derived equations and inequations.

Lemma 3.1. "Simplified" Dirichlet eta function is derived directly from Dirichlet eta function with Euler formula application and it will intrinsically incorporate actual location [but not actual positions] of all nontrivial zeros.

Proof. Denote complex number ($\mathbb{C}$) as $z = x + \imath \cdot y$. Then $z = \text{Re}(z) + \imath \cdot \text{Im}(z)$ with Re(z) = x and Im(z) = y; modulus of z, $|z| = \sqrt{Re(z)^2 + Im(z)^2} = \sqrt{x^2 + y^2}$; and $|z|^2 = x^2 + y^2$.

Euler formula is commonly stated as $e^{\imath x} = \cos x + \imath \cdot \sin x$. Euler identity (where $x = \pi$) is $e^{\imath \pi} = \cos \pi + \imath \cdot \sin \pi = $ - 1 + 0 [or stated as $e^{\imath \pi} + 1 = 0$]. The n^s of $\zeta(s)$ is expanded to $n^s = n^{(\sigma+\imath t)} = n^\sigma e^{t\ln(n)\cdot\imath}$ since $n^t = e^{t\ln(n)}$. Apply Euler formula to n^s result in $n^s = n^\sigma(\cos(t\ln(n)) + \imath\cdot\sin(t\ln(n)))$. This is written in trigonometric form [designated by short-hand notation $n^s(Euler)$] whereby n^σ is modulus and $t\ln(n)$ is polar angle (argument).

Apply $n^s(Euler)$ to Eq. (1). Then $\zeta(s) = \text{Re}\{\zeta(s)\} + \imath\cdot\text{Im}\{\zeta(s)\}$ with
$Re\{\zeta(s)\} = \sum_{n=1}^{\infty} n^{-\sigma} \cos(t\ln(n))$ and $Im\{\zeta(s)\} = \sum_{n=1}^{\infty} n^{-\sigma} \sin(t\ln(n))$. As Eq. (1) is defined only for $\sigma > 1$ where zeros never occur, we will not carry out further treatment here.

Apply $n^s(Euler)$ to Eq. (3). Then $\zeta(s) = \gamma \cdot \eta(s) = \gamma \cdot [Re\{\eta(s)\} + \imath\cdot\text{Im}\{\eta(s)\}]$ with
$$Re\{\eta(s)\} = \sum_{n=1}^{\infty} ((2n-1)^{-\sigma} \cos(t\ln(2n-1)) - (2n)^{-\sigma} \cos(t\ln(2n)));$$
$$Im\{\eta(s)\} = \sum_{n=1}^{\infty} ((2n)^{-\sigma} \sin(t\ln(2n)) - (2n-1)^{-\sigma} \sin(t\ln(2n-1)));$$
and proportionality factor $\gamma = \dfrac{1}{(1 - 2^{1-s})}$.

Complex number s in critical strip is designated by $s = \sigma + \imath t$ for $0 < t < +\infty$ and $s = \sigma - \imath t$ for $-\infty < t < 0$. Nontrivial zeros equating to $\zeta(s) = 0$ give rise to our desired $\eta(s) = 0$. Modulus of $\eta(s)$, $|\eta(s)|$, is defined as $\sqrt{(Re\{\eta(s)\})^2 + (Im\{\eta(s)\})^2}$ with $|\eta(s)|^2 = (Re\{\eta(s)\})^2 + (Im\{\eta(s)\})^2$. Mathematically $|\eta(s)| = |\eta(s)|^2 = 0$ is an unique condition giving rise to $\eta(s) = 0$ occurring only when $Re\{\eta(s)\} = Im\{\eta(s)\} = 0$ as any non-zero values for $Re\{\eta(s)\}$ and/or $Im\{\eta(s)\}$ will always result in $|\eta(s)|$ and $|\eta(s)|^2$ having non-zero values. Important implication is that sum of $Re\{\eta(s)\}$ and $Im\{\eta(s)\}$ equating to zero [given by Eq. (4)] must always hold when $|\eta(s)| = |\eta(s)|^2 = 0$ and consequently $\eta(s) = 0$.

$$\sum ReIm\{\eta(s)\} = Re\{\eta(s)\} + Im\{\eta(s)\} = 0 \tag{4}$$

In principle, advocating for existence of theoretical s values leading to non-zero values in $Re\{\eta(s)\}$ and $Im\{\eta(s)\}$ depicted as possibility $+Re\{\eta(s)\} = -Im\{\eta(s)\}$ or $-Re\{\eta(s)\} = +Im\{\eta(s)\}$ could satisfy Eq. (4). This reverse implication is not necessarily true as these s values will not result in $|\eta(s)| = |\eta(s)|^2 = 0$. In any event, we need not consider these two possibilities since solving Riemann hypothesis involves nontrivial zeros defined by $\eta(s) = 0$ with non-zero values in $Re\{\eta(s)\}$ and/or $Im\{\eta(s)\}$ being not compatible with $\eta(s) = 0$.

Riemann hypothesis proposed all nontrivial zeros to be located on critical line. This location is conjectured to be uniquely associated with presence of exact DA homogeneity in derived equation and inequation of Dirichlet Sigma-Power Law with Eq. (4) intrinsically incorporated into this Law as the $\eta(s) = 0$ definition for nontrivial zeros equates to Eq. (4).

Apply trigonometry identity $\cos(x) - \sin(x) = \sqrt{2}\sin\left(x + \dfrac{3}{4}\pi\right)$ to $Re\{\eta(s)\} + Im\{\eta(s)\}$ to get Eq. (5) with terms in last line built by mixture of terms from $Re\{\eta(s)\}$ and $Im\{\eta(s)\}$.

$$\sum ReIm\{\eta(s)\} = \sum_{n=1}^{\infty} [(2n-1)^{-\sigma} \cos(t\ln(2n-1)) - (2n-1)^{-\sigma} \sin(t\ln(2n-1))$$
$$- (2n)^{-\sigma} \cos(t\ln(2n)) + (2n)^{-\sigma} \sin(t\ln(2n))]$$
$$= \sum_{n=1}^{\infty} [(2n-1)^{-\sigma} \sqrt{2}\sin(t\ln(2n-1) + \dfrac{3}{4}\pi) - (2n)^{-\sigma}\sqrt{2}\sin(t\ln(2n) + \dfrac{3}{4}\pi)] \tag{5}$$

for $-\infty < \sigma < \infty$. It can be used to find all trivial zeros on horizontal line at $\imath t = 0$ occurring when σ = -2, -4, -6, -8, -10,..., ∞ whereby $\zeta(s) = 0$ because factor $\sin(\frac{\pi s}{2})$ vanishes. Γ is gamma function, an extension of factorial function [a product function denoted by ! notation whereby n! = $n(n-1)(n-2)\ldots(n-(n-1))$] with its argument shifted down by 1, to real and complex numbers. That is, if n is a positive integer, $\Gamma(n) = (n-1)!$

$$\zeta(s) = \frac{1}{(1-2^{1-s})} \sum_{n=1}^{\infty} \frac{(-1)^{n+1}}{n^s} \qquad (3)$$
$$= \frac{1}{(1-2^{1-s})} \left(\frac{1}{1^s} - \frac{1}{2^s} + \frac{1}{3^s} - \cdots \right)$$

Eq. (3) is defined for all $\sigma > 0$ values except for simple pole at $\sigma = 1$. As alluded to above, ζ(s) without $\frac{1}{(1-2^{1-s})}$ viz. $\sum_{n=1}^{\infty} \frac{(-1)^{n+1}}{n^s}$ is η(s). It is a holomorphic function of s defined by analytic continuation and is mathematically defined at $\sigma = 1$ whereby analogous trivial zeros with presence only for η(s) [but not for ζ(s)] on vertical straight line $\sigma = 1$ are found at $s = 1 \pm \imath \cdot \frac{2\pi k}{\ln(2)}$ where k = 1, 2, 3, 4, 5, ..., ∞.

Figure 2 above depict complex variable s (= $\sigma \pm \imath t$) as INPUT with x-axis denoting real part Re{s} equating to σ, and y-axis denoting imaginary part Im{s} equating to t. Figures 3, 4 and 5 above respectively depict $\zeta(s)$ as OUTPUT for real values of t running from 0 to 34 at $\sigma = \frac{1}{2}$ (critical line), $\sigma = \frac{2}{5}$ (non-critical line), and $\sigma = \frac{3}{5}$ (non-critical line) with x-axis denoting real part Re{$\zeta(s)$} and y-axis denoting imaginary part Im{$\zeta(s)$}. There are infinite types-of-spirals possibilities associated with each σ value arising from all infinite σ values in critical strip. Mathematically proving all nontrivial zeros location on critical line as denoted by solitary $\sigma = \frac{1}{2}$ value equates to geometrically proving all Origin intercepts occurrence at solitary $\sigma = \frac{1}{2}$ value. Both result in rigorous proof for Riemann hypothesis.

3. Prerequisite lemma, corollary and propositions for Riemann hypothesis

Original equation $\eta(s)$, *proxy* for $\zeta(s)$, is treated as unique mathematical object with key properties and behaviors. Containing all x-axis, y-axis and Origin intercepts, it will intrinsically incorporate *actual location [but not actual positions]* of all Gram[y=0] points, Gram[x=0] points and nontrivial zeros. Proofs on lemma, corollary and propositions on nontrivial zeros depict exact and inexact DA homogeneity in both derived equation and inequation. Parallel procedure on Gram[y=0] and Gram[x=0] points below depict exact and inexact DA homogeneity in similarly derived equations and inequations.

Lemma 3.1. "Simplified" Dirichlet eta function is derived directly from Dirichlet eta function with Euler formula application and it will intrinsically incorporate actual location [but not actual positions] of all nontrivial zeros.

Proof. Denote complex number ($\mathbb{C}$) as z = x + $\imath \cdot$y. Then z = Re(z) + $\imath \cdot$Im(z) with Re(z) = x and Im(z) = y; modulus of z, $|z| = \sqrt{Re(z)^2 + Im(z)^2} = \sqrt{x^2 + y^2}$; and $|z|^2 = x^2 + y^2$.

Euler formula is commonly stated as $e^{\imath x} = \cos x + \imath \cdot \sin x$. Euler identity (where $x = \pi$) is $e^{\imath \pi} = \cos \pi + \imath \cdot \sin \pi = -1 + 0$ [or stated as $e^{\imath \pi} + 1 = 0$]. The n^s of ζ(s) is expanded

Book 2 Three Open Problems by Riemann and Polignac

to $n^s = n^{(\sigma+it)} = n^\sigma e^{t\ln(n)\cdot i}$ since $n^t = e^{t\ln(n)}$. Apply Euler formula to n^s result in $n^s = n^\sigma(\cos(t\ln(n)) + i\cdot\sin(t\ln(n)))$. This is written in trigonometric form [designated by short-hand notation $n^s(Euler)$] whereby n^σ is modulus and $t\ln(n)$ is polar angle (argument).

Apply $n^s(Euler)$ to Eq. (1). Then $\zeta(s) = Re\{\zeta(s)\} + i\cdot Im\{\zeta(s)\}$ with
$Re\{\zeta(s)\} = \sum_{n=1}^{\infty} n^{-\sigma}\cos(t\ln(n))$ and $Im\{\zeta(s)\} = \sum_{n=1}^{\infty} n^{-\sigma}\sin(t\ln(n))$. As Eq. (1) is defined only for $\sigma > 1$ where zeros never occur, we will not carry out further treatment here.

Apply $n^s(Euler)$ to Eq. (3). Then $\zeta(s) = \gamma\cdot\eta(s) = \gamma\cdot[Re\{\eta(s)\} + i\cdot Im\{\eta(s)\}]$ with
$$Re\{\eta(s)\} = \sum_{n=1}^{\infty}((2n-1)^{-\sigma}\cos(t\ln(2n-1)) - (2n)^{-\sigma}\cos(t\ln(2n)));$$
$$Im\{\eta(s)\} = \sum_{n=1}^{\infty}((2n)^{-\sigma}\sin(t\ln(2n)) - (2n-1)^{-\sigma}\sin(t\ln(2n-1)));$$
and proportionality factor $\gamma = \dfrac{1}{(1-2^{1-s})}$.

Complex number s in critical strip is designated by $s = \sigma + it$ for $0 < t < +\infty$ and $s = \sigma - it$ for $-\infty < t < 0$. Nontrivial zeros equating to $\zeta(s) = 0$ give rise to our desired $\eta(s) = 0$. Modulus of $\eta(s)$, $|\eta(s)|$, is defined as $\sqrt{(Re\{\eta(s)\})^2 + (Im\{\eta(s)\})^2}$ with $|\eta(s)|^2 = (Re\{\eta(s)\})^2 + (Im\{\eta(s)\})^2$. Mathematically $|\eta(s)| = |\eta(s)|^2 = 0$ is an unique condition giving rise to $\eta(s) = 0$ occurring only when $Re\{\eta(s)\} = Im\{\eta(s)\} = 0$ as any non-zero values for $Re\{\eta(s)\}$ and/or $Im\{\eta(s)\}$ will always result in $|\eta(s)|$ and $|\eta(s)|^2$ having non-zero values. Important implication is that sum of $Re\{\eta(s)\}$ and $Im\{\eta(s)\}$ equating to zero [given by Eq. (4)] must always hold when $|\eta(s)| = |\eta(s)|^2 = 0$ and consequently $\eta(s) = 0$.

$$\sum ReIm\{\eta(s)\} = Re\{\eta(s)\} + Im\{\eta(s)\} = 0 \qquad (4)$$

In principle, advocating for existence of theoretical s values leading to non-zero values in $Re\{\eta(s)\}$ and $Im\{\eta(s)\}$ depicted as possibility $+Re\{\eta(s)\} = -Im\{\eta(s)\}$ or $-Re\{\eta(s)\} = +Im\{\eta(s)\}$ could satisfy Eq. (4). This reverse implication is not necessarily true as these s values will not result in $|\eta(s)| = |\eta(s)|^2 = 0$. In any event, we need not consider these two possibilities since solving Riemann hypothesis involves nontrivial zeros defined by $\eta(s) = 0$ with non-zero values in $Re\{\eta(s)\}$ and/or $Im\{\eta(s)\}$ being not compatible with $\eta(s) = 0$.

Riemann hypothesis proposed all nontrivial zeros to be located on critical line. This location is conjectured to be uniquely associated with presence of exact DA homogeneity in derived equation and inequation of Dirichlet Sigma-Power Law with Eq. (4) intrinsically incorporated into this Law as the $\eta(s) = 0$ definition for nontrivial zeros equates to Eq. (4).

Apply trigonometry identity $\cos(x) - \sin(x) = \sqrt{2}\sin\left(x + \dfrac{3}{4}\pi\right)$ to $Re\{\eta(s)\} + Im\{\eta(s)\}$ to get Eq. (5) with terms in last line built by mixture of terms from $Re\{\eta(s)\}$ and $Im\{\eta(s)\}$.

$$\sum ReIm\{\eta(s)\} = \sum_{n=1}^{\infty}[(2n-1)^{-\sigma}\cos(t\ln(2n-1)) - (2n-1)^{-\sigma}\sin(t\ln(2n-1))$$
$$- (2n)^{-\sigma}\cos(t\ln(2n)) + (2n)^{-\sigma}\sin(t\ln(2n))]$$
$$= \sum_{n=1}^{\infty}[(2n-1)^{-\sigma}\sqrt{2}\sin(t\ln(2n-1) + \tfrac{3}{4}\pi) - (2n)^{-\sigma}\sqrt{2}\sin(t\ln(2n) + \tfrac{3}{4}\pi)] \qquad (5)$$

When depicted in terms of Eq. (4), Eq. (5) becomes

$$\sum_{n=1}^{\infty}(2n)^{-\sigma}\sqrt{2}\sin(t\ln(2n)+\frac{3}{4}\pi) = \sum_{n=1}^{\infty}(2n-1)^{-\sigma}\sqrt{2}\sin(t\ln(2n-1)+\frac{3}{4}\pi)$$

$$\sum_{n=1}^{\infty}(2n)^{-\sigma}\sqrt{2}\sin(t\ln(2n)+\frac{3}{4}\pi) - \sum_{n=1}^{\infty}(2n-1)^{-\sigma}\sqrt{2}\sin(t\ln(2n-1)+\frac{3}{4}\pi) = 0 \qquad (6)$$

Eq. (6) in discrete (summation) format is a non-Hybrid integer sequence equation – see Appendix C. $\eta(s)$ calculations for all σ values result in infinitely many non-Hybrid integer sequence equations for $0<\sigma<1$ critical strip region of interest with n = 1, 2, 3, 4, 5,..., ∞ as discrete integer number values, or n = 1 to ∞ as continuous real numbers values with Riemann integral application. These equations will geometrically represent entire plane of critical strip, thus (at least) allowing our proposed proof to be of a complete nature.

Eq. (6) being the "simplified" Dirichlet eta function derived directly from $\eta(s)$ will intrinsically incorporate *actual location [but not actual positions]* of all nontrivial zeros. *The proof is now complete for Lemma 3.1*□.

Proposition 3.2. Dirichlet Sigma-Power Law in continuous (integral) format given as equation and inequation can both be derived directly from "simplified" Dirichlet eta function in discrete (summation) format with Riemann integral application. [Note: Dirichlet Sigma-Power Law in continuous (integral) format here refers to the end-product obtained from "first key step of converting Riemann zeta function into its continuous format version".]

Proof. In Calculus, integration is reverse process of differentiation viewed geometrically as numerical "total area value" solution enclosed by curve of function and x-axis. Apply definite integral I between limits (or points) a and b is to compute its value when $\Delta x \longrightarrow 0$, i.e. $I = \lim_{\Delta x \to 0}\sum_{i=1}^{n}f(x_i)\Delta x_i = \int_{a}^{b}f(x)dx$. This is Riemann integral of function f(x) in interval [a, b] where a<b. Apply Riemann integral to "simplified" Dirichlet eta function in [$\Delta x \longrightarrow 1$] discrete (summation) format which intrinsically incorporates *actual location [but not actual positions]* of all nontrivial zeros criterion to obtain Dirichlet Sigma-Power Law in [$\Delta x \longrightarrow 0$] continuous (integral) format with the later validly representing the former. Then Dirichlet Sigma-Power Law will also fullfil this criterion. Due to resemblance to power law functions in σ from s = $\sigma + \imath t$ being exponent of a power function n^{σ}, logarithm scale use, and harmonic $\zeta(s)$ series connection in Zipf's law; we elect to call this Law by its given name. A characteristic and crucial step of this Law is its exact formula expression in usual mathematical language [y = $f(x_1, x_2)$ format description for a 2-variable function with $(2n)$ and $(2n-1)$ parameters] consist of y = $f(t, \sigma)$ with discrete n = 1, 2, 3, 4, 5,..., ∞ or continuous n = 1 to ∞; $-\infty < t < +\infty$; and $0 < \sigma < 1$.

Note: A proper integral is a definite integral which has neither limit a or b infinite and from which the integrand does not approach infinity at any point in the range of integration. Only a proper integral will have its [solitary] combined +ve (above x-axis) and -ve (below x-axis) **non-zero** numerical "total area value" solution successfully computed from applying Riemann integral. An improper integral is a definite integral that has either or both limits a and b infinite or an integrand that approaches infinity at one or more points in the range of integration. Our resulting Dirichlet Sigma-Power Law, being improper integral (with lower limit a = 1 and upper limit b = ∞) obtained from [validly] applying Riemann integral to "simplified" Dirichlet eta function, will [expectedly] have its [multiple] +ve (above x-axis) = -ve (below x-axis) **zero** numerical "net area value" solutions successfully computed – see Propositions 3.3 and 3.4 below.

Book 2 Three Open Problems by Riemann and Polignac

With steps of manual integration shown using indefinite integrals [for simplicity], we solve definite integral below based on numerator portion of R1 with $(2n)$ parameter in Eq. (6):

$$\int_1^\infty \frac{2^{\frac{1}{2}-\sigma} \sin\left(t\ln(2n) + \frac{3\pi}{4}\right)}{n^\sigma} dn = \int_1^\infty -\frac{\sin(t\ln(2n)) - \cos(t\ln(2n))}{2^\sigma n^\sigma} dn.$$ We deduce most other important integrals to be "variations" of this particular integral containing (i) deletion of $(2n)^{-\sigma}$, $\sqrt{2}$ or $\frac{3}{4}\pi$ terms, and/or (ii) interchange of sine and cosine function. We check all derived antiderivatives to be correct using computer algebra system Maxima.

Simplifying and applying linearity, we obtain $2^{\frac{1}{2}-\sigma} \int \frac{\sin\left(t\ln(2n) + \frac{3\pi}{4}\right)}{n^\sigma} dn$.

Now solving $\int \frac{\sin\left(t\ln(2n) + \frac{3\pi}{4}\right)}{n^\sigma} dn$. Substitute $u = t\ln(2n) + \frac{3\pi}{4} \longrightarrow dn = \frac{n}{t} du$,

use $n^{1-\sigma} = e^{\frac{(1-\sigma)\left(u - t\ln(2) - \frac{3\pi}{4}\right)}{t}} = \frac{e^{\frac{(\sigma-1)(4t\ln(2) + 3\pi)}{4t}}}{t} \int e^{\frac{(1-\sigma)u}{t}} \sin(u) \, du.$

Now solving $\int e^{\frac{(1-\sigma)u}{t}} \sin(u) \, du$. We integrate by parts twice in a row: $\int \mathbf{f} g' = \mathbf{f} g - \int \mathbf{f}' g$.

First time: $\mathbf{f} = \sin(u), g' = e^{\frac{(1-\sigma)u}{t}}$

Then $\mathbf{f}' = \boxed{\cos(u)}, g = \boxed{\frac{(1-\sigma) t e^{\frac{(1-\sigma)u}{t}}}{\sigma^2 - 2\sigma + 1}}$:

$= \frac{(1-\sigma) t e^{\frac{(1-\sigma)u}{t}} \sin(u)}{\sigma^2 - 2\sigma + 1} - \int \frac{(1-\sigma) t e^{\frac{(1-\sigma)u}{t}} \cos(u)}{\sigma^2 - 2\sigma + 1} du$

Second time: $\mathbf{f} = \boxed{\cos(u)}, g' = \boxed{\frac{(1-\sigma) t e^{\frac{(1-\sigma)u}{t}}}{\sigma^2 - 2\sigma + 1}}$

Then $\mathbf{f}' = -\sin(u), g = \frac{t^2 e^{\frac{(1-\sigma)u}{t}}}{\sigma^2 - 2\sigma + 1}$:

$= \frac{(1-\sigma) t e^{\frac{(1-\sigma)u}{t}} \sin(u)}{\sigma^2 - 2\sigma + 1} - \left(\frac{t^2 e^{\frac{(1-\sigma)u}{t}} \cos(u)}{\sigma^2 - 2\sigma + 1} - \int -\frac{t^2 e^{\frac{(1-\sigma)u}{t}} \sin(u)}{\sigma^2 - 2\sigma + 1} du \right)$

Apply linearity:

$= \frac{(1-\sigma) t e^{\frac{(1-\sigma)u}{t}} \sin(u)}{\sigma^2 - 2\sigma + 1} - \left(\frac{t^2 e^{\frac{(1-\sigma)u}{t}} \cos(u)}{\sigma^2 - 2\sigma + 1} + \frac{t^2}{\sigma^2 - 2\sigma + 1} \int e^{\frac{(1-\sigma)u}{t}} \sin(u) du \right)$

As integral $\int e^{\frac{(1-\sigma)u}{t}} \sin(u) \, du$ appears again on Right Hand Side, we solve for it:

$= \frac{\frac{(1-\sigma) e^{\frac{(1-\sigma)u}{t}} \sin(u)}{t} - e^{\frac{(1-\sigma)u}{t}} \cos(u)}{\frac{\sigma^2 - 2\sigma + 1}{t^2} + 1}$

Plug in solved integrals: $\frac{e^{\frac{(\sigma-1)(4t\ln(2) + 3\pi)}{4t}}}{t} \int e^{\frac{(1-\sigma)u}{t}} \sin(u) \, du$

$= \frac{e^{\frac{(\sigma-1)(4t\ln(2) + 3\pi)}{4t}} \left(\frac{(1-\sigma) e^{\frac{(1-\sigma)u}{t}} \sin(u)}{t} - e^{\frac{(1-\sigma)u}{t}} \cos(u) \right)}{\left(\frac{\sigma^2 - 2\sigma + 1}{t^2} + 1 \right) t}$

Undo substitution $u = t\ln(2n) + \frac{3\pi}{4}$ and simplifying:

$$\left[\frac{(2n-1)\left((t-1)\sin(t\ln(2n-1))+(t+1)\cos(t\ln(2n-1))\right)}{2(t^2+1)}+C\right]_1^\infty$$

$$\int_1^\infty (2n)^\sigma\,dn = \left[\frac{(2n)^{\sigma+1}}{2(\sigma+1)}+C\right]_1^\infty \text{ and } \int_1^\infty (2n-1)^\sigma\,dn = \left[\frac{(2n-1)^{\sigma+1}}{2(\sigma+1)}+C\right]_1^\infty$$

Dirichlet Sigma-Power Law as inequation derived from Eq. (10) is given by:

$$\left[\frac{(2n)\left((t-1)\sin(t\ln(2n))+(t+1)\cos(t\ln(2n))\right)}{(2n-1)\left((t-1)\sin(t\ln(2n-1))+(t+1)\cos(t\ln(2n-1))\right)} - \frac{(2n)^{\sigma+1}}{(2n-1)^{\sigma+1}}\right]_1^\infty \neq 0 \quad (11)$$

Intended derivation of Dirichlet Sigma-Power Law as equation and inequation have been successful. *The proof is now complete for Proposition 3.2*□.

Proposition 3.3. Exact Dimensional analysis homogeneity at $\sigma = \frac{1}{2}$ in Dirichlet Sigma-Power Law as equation and inequation is (respectively) indicated by $\sum$(all fractional exponents) = whole number '1' and '3'.

Proof. Dirichlet Sigma-Power Law as equation for $\sigma = \frac{1}{2}$ value is given by:

$$\frac{1}{2t^2+\frac{1}{2}} \cdot [(2n)^{\frac{1}{2}}\left((t-\frac{1}{2})\sin(t\ln(2n))+(t+\frac{1}{2})\cos(t\ln(2n))\right) -$$

$$(2n-1)^{\frac{1}{2}}\left((t-\frac{1}{2})\sin(t\ln(2n-1))+(t+\frac{1}{2})\cos(t\ln(2n-1))\right)]_1^\infty = 0 \quad (12)$$

Respectively evaluation of definite integrals Eq. (12), Eq. (24) and Eq. (26) using limit as n → +∞ for $0 < t < +\infty$ enable countless computations resulting in t values for CIS of nontrivial zeros, Gram[y=0] and Gram[x=0] points. We evaluate Eq. (12) to obtain its expanded antiderivative:

$$\frac{1}{2t^2+\frac{1}{2}} \cdot [(2\infty)^{\frac{1}{2}}\left((t-\frac{1}{2})\sin(t\ln(2\infty))+(t+\frac{1}{2})\cos(t\ln(2\infty))\right) -$$

$$(2\infty-1)^{\frac{1}{2}}\left((t-\frac{1}{2})\sin(t\ln(2\infty-1))+(t+\frac{1}{2})\cos(t\ln(2\infty-1))\right)$$

$$-(2)^{\frac{1}{2}}\left((t-\frac{1}{2})\sin(t\ln(2))+(t+\frac{1}{2})\cos(t\ln(2))\right) +$$

$$(1)^{\frac{1}{2}}\left((t-\frac{1}{2})\sin(t\ln(1))+(t+\frac{1}{2})\cos(t\ln(1))\right)] = 0$$

$$\frac{1}{2t^2+\frac{1}{2}} \cdot [(2\infty)^{\frac{1}{2}}\left((t-\frac{1}{2})\sin(t\ln(2\infty))+(t+\frac{1}{2})\cos(t\ln(2\infty))\right) -$$

$$(2\infty-1)^{\frac{1}{2}}\left((t-\frac{1}{2})\sin(t\ln(2\infty-1))+(t+\frac{1}{2})\cos(t\ln(2\infty-1))\right)$$

$$-(2)^{\frac{1}{2}}\left((t-\frac{1}{2})\sin(t\ln(2))+(t+\frac{1}{2})\cos(t\ln(2))\right) + \left(t+\frac{1}{2}\right)] = 0$$

This is simplified to

$$\frac{1}{2t^2+\frac{1}{2}} \cdot [-(2)^{\frac{1}{2}}\left((t-\frac{1}{2})\sin(t\ln(2))+(t+\frac{1}{2})\cos(t\ln(2))\right) + \left(t+\frac{1}{2}\right)] = 0$$

Equivalent evaluation on Eq. (12) to obtain its expanded antiderivative depicted as linear combination of sine and cosine waves: $a\sin x + b\cos x = c\sin(x+\varphi)$ with $c = \sqrt{a^2+b^2}$ and $\varphi = \text{atan2}(b,a) = \tan^{-1}(\frac{b}{a})$ for a>0:

$$\frac{1}{2t^2+\frac{1}{2}} \cdot [((2\infty)(2t^2+\frac{1}{2}))^{\frac{1}{2}}\sin\left((t\ln 2\infty) + \tan^{-1}(\frac{t+\frac{1}{2}}{t-\frac{1}{2}})\right)$$

$$-((2\infty-1)(2t^2+\frac{1}{2}))^{\frac{1}{2}}\sin\left((t\ln 2\infty - 1) + \tan^{-1}(\frac{t+\frac{1}{2}}{t-\frac{1}{2}})\right)$$

$$-((2)(2t^2+\frac{1}{2}))^{\frac{1}{2}}\sin\left((t\ln 2) + \tan^{-1}(\frac{t+\frac{1}{2}}{t-\frac{1}{2}})\right) + (2t^2+\frac{1}{2})^{\frac{1}{2}}\sin\left(\tan^{-1}(\frac{t+\frac{1}{2}}{t-\frac{1}{2}})\right)] = 0$$

$$= \frac{e^{\frac{(\sigma-1)(4t\ln(2)+3\pi)}{4t}} \left((1-\sigma)e^{\frac{(1-\sigma)\left(t\ln(2n)+\frac{3\pi}{4}\right)}{t}} \frac{\sin\left(t\ln(2n)+\frac{3\pi}{4}\right)}{t} - e^{\frac{(1-\sigma)\left(t\ln(2n)+\frac{3\pi}{4}\right)}{t}} \cos\left(t\ln(2n)+\frac{3\pi}{4}\right) \right)}{\left(\frac{\sigma^2-2\sigma+1}{t^2}+1\right)t}$$

Plug in solved integrals: $2^{\frac{1}{2}-\sigma} \int \frac{\sin\left(t\ln(2n)+\frac{3\pi}{4}\right)}{n^\sigma} dn$

$$= \frac{2^{\frac{1}{2}-\sigma} e^{\frac{(\sigma-1)(4t\ln(2)+3\pi)}{4t}} \left((1-\sigma)e^{\frac{(1-\sigma)\left(t\ln(2n)+\frac{3\pi}{4}\right)}{t}} \frac{\sin\left(t\ln(2n)+\frac{3\pi}{4}\right)}{t} - e^{\frac{(1-\sigma)\left(t\ln(2n)+\frac{3\pi}{4}\right)}{t}} \cos\left(t\ln(2n)+\frac{3\pi}{4}\right) \right)}{\left(\frac{\sigma^2-2\sigma+1}{t^2}+1\right)t}$$

By rewriting and simplifying, $\int_1^\infty \frac{2^{\frac{1}{2}-\sigma} \sin\left(t\ln(2n)+\frac{3\pi}{4}\right)}{n^\sigma} dn$ is finally solved as

$$\left[\frac{(2n)^{1-\sigma}\left((t+\sigma-1)\sin(t\ln(2n)) + (t-\sigma+1)\cos(t\ln(2n))\right)}{2\left(t^2+(\sigma-1)^2\right)} + C \right]_1^\infty \quad (7)$$

For denominator portion of R1 with $(2n-1)$ parameter in Eq. (6), Eq. (7) equates to

$$\left[\frac{(2n-1)^{1-\sigma}\left((t+\sigma-1)\sin(t\ln(2n-1)) + (t-\sigma+1)\cos(t\ln(2n-1))\right)}{2\left(t^2+(\sigma-1)^2\right)} + C \right]_1^\infty \quad (8)$$

Dirichlet Sigma-Power Law as equation derived from Eq. (6) is given by:

$$\frac{1}{2(t^2+(\sigma-1)^2)} \cdot [(2n)^{1-\sigma}\left((t+\sigma-1)\sin(t\ln(2n)) + (t-\sigma+1)\cos(t\ln(2n))\right) -$$

$$(2n-1)^{1-\sigma}\left((t+\sigma-1)\sin(t\ln(2n-1)) + (t-\sigma+1)\cos(t\ln(2n-1))\right)]_1^\infty = 0 \quad (9)$$

Apply Ratio Study to Eq. (6) – see Appendix B. This involves [intentional] incorrect but "balanced" rearrangement of terms in Eq. (6) giving rise to Eq. (10) which is a non-Hybrid integer sequence inequation. Left-hand side contains 'cyclical' sine function in first term (Ratio R1) and 'non-cyclical' power function in second term (Ratio R2).

$$\frac{\sum_{n=1}^\infty \sqrt{2}\sin(t\ln(2n)+\frac{3}{4}\pi)}{\sum_{n=1}^\infty \sqrt{2}\sin(t\ln(2n-1)+\frac{3}{4}\pi)} - \frac{\sum_{n=1}^\infty (2n)^\sigma}{\sum_{n=1}^\infty (2n-1)^\sigma} \neq 0 \quad (10)$$

Apply Riemann integral to selected parts of Eq. (10) without depicting steps of calculation:
$$\int_1^\infty \sqrt{2}\sin\left(t\ln(2n)+\frac{3\pi}{4}\right) dn =$$
$$\left[\frac{(2n)\left((t-1)\sin(t\ln(2n)) + (t+1)\cos(t\ln(2n))\right)}{2(t^2+1)} + C \right]_1^\infty$$
and $\int_1^\infty \sqrt{2}\sin\left(t\ln(2n-1)+\frac{3\pi}{4}\right) dn =$

This is simplified to

$$\frac{1}{2t^2 + \frac{1}{2}} \cdot \left[-((2)(2t^2 + \frac{1}{2}))^{\frac{1}{2}} \sin\left((t \ln 2) + \tan^{-1}(\frac{t + \frac{1}{2}}{t - \frac{1}{2}}) \right) + (2t^2 + \frac{1}{2})^{\frac{1}{2}} \sin\left(\tan^{-1}(\frac{t + \frac{1}{2}}{t - \frac{1}{2}}) \right) \right] = 0$$

Relevant t values for all nontrivial zeros at $\sigma = \frac{1}{2}$ plotted against the expanded antiderivative

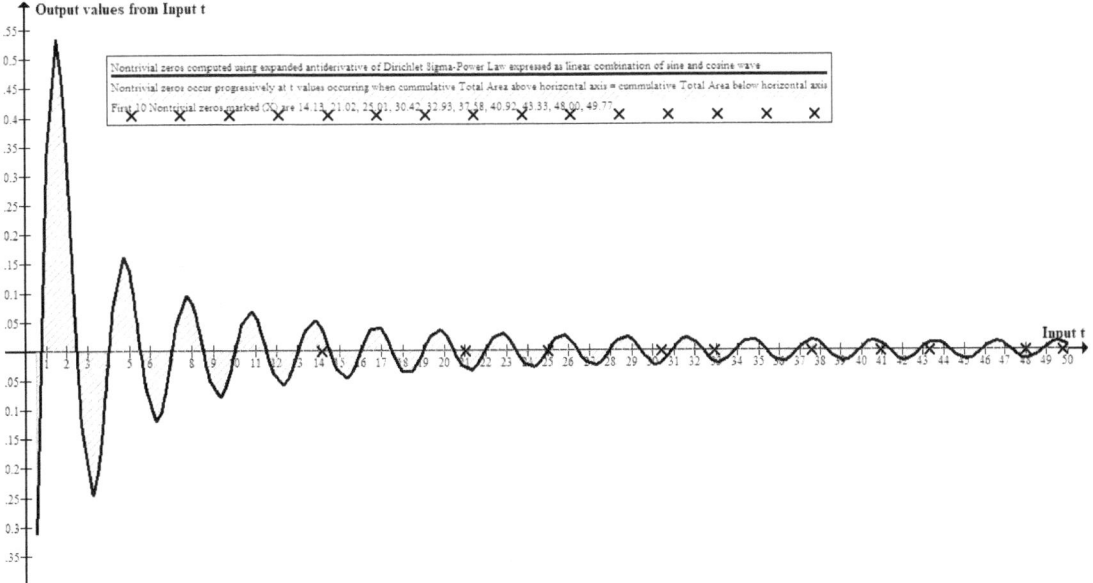

FIGURE 6. First 10 nontrivial zeros at $\sigma = \frac{1}{2}$ displayed using Dirichlet Sigma-Power Law.

depicted as linear combination of sine and cosine waves is shown in Figure 6. The phenomenon of "monotonously decreasing waves" with line of symmetry being *uniquely* the horizontal axis will only occur when $\sigma = \frac{1}{2}$.

Dirichlet Sigma-Power Law as inequation for $\sigma = \frac{1}{2}$ value is given by:

$$\left[\frac{(2n)\left((t-1)\sin(t\ln(2n)) + (t+1)\cos(t\ln(2n))\right)}{(2n-1)\left((t-1)\sin(t\ln(2n-1)) + (t+1)\cos(t\ln(2n-1))\right)} - \frac{(2n)^{\frac{3}{2}}}{(2n-1)^{\frac{3}{2}}} \right]_1^\infty \neq 0 \quad (13)$$

$\sum$(all fractional exponents) as $2(1-\sigma)$ = whole number '1' for Eq. (12) and $2(\sigma+1)$ = whole number '3' for Eq. (13). These findings signify presence of complete set nontrivial zeros for Eq. (12) and Eq. (13). *The proof is now complete for Proposition 3.3*□.

Corollary 3.4. Inexact Dimensional analysis homogeneity at $\sigma \neq \frac{1}{2}$ [illustrated using $\sigma = \frac{2}{5}$] in Dirichlet Sigma-Power Law as equation and inequation is (respectively) indicated by $\sum$(all fractional exponents) = fractional number '$\neq 1$' and '$\neq 3$'.

Proof. Dirichlet Sigma-Power Law as equation for $\sigma = \frac{2}{5}$ value is given by:

$$\frac{1}{2t^2 + \frac{18}{25}} \cdot \left[(2n)^{\frac{3}{5}} \left((t - \frac{3}{5}) \sin(t \ln(2n)) + (t + \frac{3}{5}) \cos(t \ln(2n)) \right) - \right.$$

$$\left. (2n-1)^{\frac{3}{5}} \left((t - \frac{3}{5}) \sin(t \ln(2n-1)) + (t + \frac{3}{5}) \cos(t \ln(2n-1)) \right) \right]_1^\infty = 0 \quad (14)$$

We evaluate Eq. (14) to obtain its expanded antiderivative:

Book 2 Three Open Problems by Riemann and Polignac

$\frac{1}{2t^2+\frac{18}{25}} \cdot [(2\infty)^{\frac{3}{5}} \left((t-\frac{3}{5})\sin(t\ln(2\infty)) + (t+\frac{3}{5})\cos(t\ln(2\infty))\right) -$

$(2\infty-1)^{\frac{3}{5}}\left((t-\frac{3}{5})\sin(t\ln(2\infty-1)) + (t+\frac{3}{5})\cos(t\ln(2\infty-1))\right)$

$-(2)^{\frac{3}{5}}\left((t-\frac{3}{5})\sin(t\ln(2)) + (t+\frac{3}{5})\cos(t\ln(2))\right) +$

$(1)^{\frac{3}{5}}\left((t-\frac{3}{5})\sin(t\ln(1)) + (t+\frac{3}{5})\cos(t\ln(1))\right)] = 0$

$\frac{1}{2t^2+\frac{18}{25}} \cdot [(2\infty)^{\frac{3}{5}} \left((t-\frac{3}{5})\sin(t\ln(2\infty)) + (t+\frac{3}{5})\cos(t\ln(2\infty))\right) -$

$(2\infty-1)^{\frac{3}{5}}\left((t-\frac{3}{5})\sin(t\ln(2\infty-1)) + (t+\frac{3}{5})\cos(t\ln(2\infty-1))\right)$

$-(2)^{\frac{3}{5}}\left((t-\frac{3}{5})\sin(t\ln(2)) + (t+\frac{3}{5})\cos(t\ln(2))\right) + (t+\frac{3}{5})] = 0$

This is simplified to

$\frac{1}{2t^2+\frac{18}{25}} \cdot [-(2)^{\frac{3}{5}}\left((t-\frac{3}{5})\sin(t\ln(2)) + (t+\frac{3}{5})\cos(t\ln(2))\right) + (t+\frac{3}{5})] = 0$

Equivalent valuation on Eq. (14) to obtain its expanded antiderivative depicted as linear combination of sine and cosine waves: $a\sin x + b\cos x = c\sin(x+\varphi)$ with $c = \sqrt{a^2+b^2}$ and $\varphi = \text{atan2}(b,a) = \tan^{-1}(\frac{b}{a})$ for $a>0$:

$\frac{1}{2t^2+\frac{18}{25}} \cdot [((2\infty)(2t^2+\frac{18}{25}))^{\frac{3}{5}} \sin\left((t\ln 2\infty) + \tan^{-1}(\frac{t+\frac{3}{5}}{t-\frac{3}{5}})\right)$

$-((2\infty-1)(2t^2+\frac{18}{25}))^{\frac{3}{5}} \sin\left((t\ln 2\infty - 1) + \tan^{-1}(\frac{t+\frac{3}{5}}{t-\frac{3}{5}})\right)$

$-((2)(2t^2+\frac{18}{25}))^{\frac{3}{5}} \sin\left((t\ln 2) + \tan^{-1}(\frac{t+\frac{3}{5}}{t-\frac{3}{5}})\right) + (2t^2+\frac{18}{25})^{\frac{3}{5}} \sin\left(\tan^{-1}(\frac{t+\frac{3}{5}}{t-\frac{3}{5}})\right)] = 0$

This is simplified to

$\frac{1}{2t^2+\frac{18}{25}} \cdot [-((2)(2t^2+\frac{18}{25}))^{\frac{3}{5}} \sin\left((t\ln 2) + \tan^{-1}(\frac{t+\frac{3}{5}}{t-\frac{3}{5}})\right) + (2t^2+\frac{18}{25})^{\frac{3}{5}} \sin\left(\tan^{-1}(\frac{t+\frac{3}{5}}{t-\frac{3}{5}})\right)] = 0$

Relevant t values [for non-existent nontrivial zeros] at $\sigma = \frac{2}{5}$ plotted against the expanded antiderivative depicted as linear combination of sine and cosine waves is shown in Figure 7. The phenomenon of "monotonously decreasing waves" with line of symmetry being *not uniquely* the horizontal axis will always occur when $\sigma \neq \frac{1}{2}$.

Dirichlet Sigma-Power Law as inequation for $\sigma = \frac{2}{5}$ value is given by:

$$\left[\frac{(2n)\left((t-1)\sin(t\ln(2n)) + (t+1)\cos(t\ln(2n))\right)}{(2n-1)\left((t-1)\sin(t\ln(2n-1))(t+1)\cos(t\ln(2n-1))\right)} - \frac{(2n)^{\frac{7}{5}}}{(2n-1)^{\frac{7}{5}}}\right]_1^\infty \neq 0 \quad (15)$$

$\sum$(all fractional exponents) as $2(1-\sigma)$ = fractional number '$\neq 1$' for Eq. (14) and $2(\sigma+1)$ = fractional number '$\neq 3$' for Eq. (15). These findings signify absence of complete set nontrivial zeros for Eq. (14) and Eq. (15). *The proof is now complete for Corollary 3.4*.□

4. Rigorous proof for Riemann hypothesis summarized as Theorem Riemann I – IV

$\zeta(s) = \frac{1}{s-1} + \frac{1}{2} + 2\int_0^\infty \frac{\sin(s\arctan t)}{(1+t^2)^{\frac{s}{2}}(e^{2\pi t}-1)} dt$ is integral relation (cf. Abel-Plana summation formula [3][4]) for all $s \in \mathbb{C}$ and $s \neq 1$. This integral is insufficient for our purpose as it involves in-

Creed Odyssey in Mathematics and Medicine series

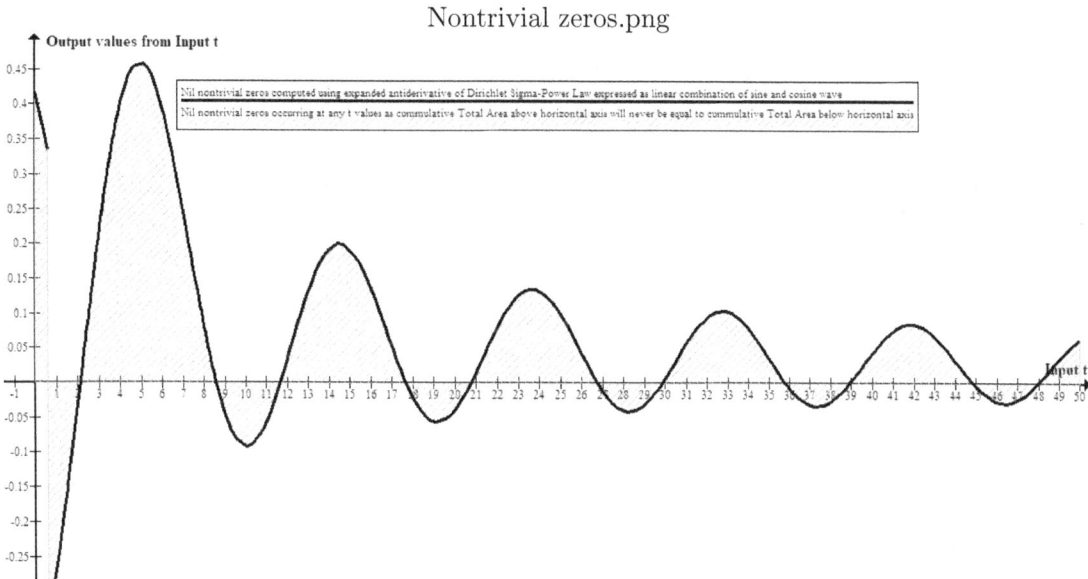

FIGURE 7. Nil nontrivial zeros present at $\sigma = \frac{2}{5}$ displayed using Dirichlet Sigma-Power Law.

tegration with respect to t [instead of n] for $\zeta(s)$ [instead of $\eta(s)$]. Rigorous proof for Riemann hypothesis is summarized by Theorem Riemann I – IV. One could obtain this proof with only using Dirichlet Sigma-Power Law [solely] as equation. For completeness and clarification of this proof, we supply following important mathematical arguments.

For $0<\sigma<1$, then $0<2(1-\sigma)<2$. The only whole number between 0 and 2 is '1' which coincide with $\sigma = \frac{1}{2}$. When $0<\sigma<\frac{1}{2}$ and $\frac{1}{2}<\sigma<1$, then $0<2(1-\sigma)<1$ and $1<2(1-\sigma)<2$.

For $0<\sigma<1$, $2<2(\sigma+1)<4$. The only whole number between 2 and 4 is '3' which coincide with $\sigma = \frac{1}{2}$. When $0<\sigma<\frac{1}{2}$ and $\frac{1}{2}<\sigma<1$, then $2<2(\sigma+1)<3$ and $3<2(\sigma+1)<4$.

Legend: **R** = all real numbers. For $0<\sigma<1$, σ consist of $0<\mathbf{R}<1$. For $0<2(1-\sigma)<2$ and $2<2(\sigma+1)<4$, $2(1-\sigma)$ and $2(\sigma+1)$ must (respectively) consist of $0<\mathbf{R}<2$ and $2<\mathbf{R}<4$. An important caveat is that previously used phrases such as "fractional exponent σ" and "$\sum$(all fractional exponents) = whole number '1' [or '3'] and fractional number '$\neq$1' [or '$\neq$3']", although not incorrect *per se*, should respectively be replaced by "real number exponent σ" and "$\sum$(all real number exponents) = whole number '1' [or '3'] and real number '$\neq$1' [or '$\neq$3'][5]" for complete accurracy. We apply this caveat to Theorem Riemann I – IV.

Footnote 5: As whole numbers $\subset$ real numbers, one could also depict this phrase as "$\sum$(all real number exponents) = real number '1' [or '3'] and real number '$\neq$1' [or '$\neq$3']".

Theorem Riemann I. Derived from *proxy* Dirichlet eta function, "simplified" Dirichlet eta function will exclusively contain *de novo* property for actual location [but not actual positions] of all nontrivial zeros.

Proof. The phrase "actual location [but not actual positions] of all nontrivial zeros" can be validly shortened to "actual location of all nontrivial zeros" as used in Theorem Riemann II, III and IV. *The proof for Theorem Riemann I is now complete as it successfully incorporates proof for Lemma 3.1*□.

Theorem Riemann II. Dirichlet Sigma-Power Law [in continuous (integral) format] as equation and inequation which are both derived from "simplified" Dirichlet eta function [in discrete (summation) format] will exclusively manifest exact DA homogeneity in equation and inequation only when real number exponent $\sigma = \frac{1}{2}$.

Proof. *The proof for Theorem Riemann II is now complete as it successfully incorporates proofs from Proposition 3.2 on derivation for equation and inequation of Dirichlet Sigma-Power Law [with both containing de novo property for "actual location of all nontrivial zeros"] and Proposition 3.3 on manifestation of exact DA homogeneity in Dirichlet Sigma-Power Law as equation and inequation when real number exponent $\sigma = \frac{1}{2}$*□.

Theorem Riemann III. Real number exponent $\sigma = \frac{1}{2}$ in Dirichlet Sigma-Power Law as equation and inequation satisfying exact DA homogeneity is identical to σ variable in Riemann hypothesis which propose σ to also have exclusive value of $\frac{1}{2}$ (representing critical line) for "actual location of all nontrivial zeros", thus fully supporting Riemann hypothesis to be true with further clarification by Theorem Riemann IV.

Proof. Since $s = \sigma \pm \imath t$, complete set of nontrivial zeros which is defined by $\eta(s) = 0$ is exclusively associated with one (and only one) particular $\eta(\sigma \pm \imath t) = 0$ value solution, and by default one (and only one) particular σ [conjecturally] $= \frac{1}{2}$ value solution. When performing exact DA homogeneity on Dirichlet Sigma-Power Law as equation and inequation [with both containing de novo property for "actual location of all nontrivial zeros"], the phrase "If real number exponent σ has exclusively $\frac{1}{2}$ value, only then will exact DA homogeneity be satisfied" implies one (and only one) possible mathematical solution. Theorem Riemann III reflect Theorem Riemann II on presence of exact DA homogeneity for $\sigma = \frac{1}{2}$ in Dirichlet Sigma-Power Law as equation and inequation. This Law has identical σ variable as that referred to by Riemann hypothesis [whereby σ here uniquely refer to critical line]. *The proof for Theorem Riemann III is now complete as it independently refers to simultaneous association of confirmed (i) solitary $\sigma = \frac{1}{2}$ value in Dirichlet Sigma-Power Law as equation and inequation satisfying exact DA homogeneity and (ii) critical line defined by solitary $\sigma = \frac{1}{2}$ value being the "actual location [but with no request to determine actual positions]" of all nontrivial zeros as proposed in original Riemann hypothesis*□.

Theorem Riemann IV. Condition 1. All $\sigma \neq \frac{1}{2}$ values (non-critical lines), viz. $0 < \sigma < \frac{1}{2}$ and $\frac{1}{2} < \sigma < 1$ values, exclusively does not contain "actual location of all nontrivial zeros" [manifesting de novo inexact DA homogeneity in equation and inequation], together with Condition 2. One (& only one) $\sigma = \frac{1}{2}$ value (critical line) exclusively contains "actual location of all nontrivial zeros" [manifesting de novo exact DA homogeneity in equation and inequation], fully support Riemann hypothesis to be true when these two mutually inclusive conditions are met.

Proof. Condition 2 Theorem Riemann IV simply reflect proof from Theorem Riemann III [incorporating Proposition 3.3] for "actual location of all nontrivial zeros" exclusively on critical line manifesting de novo exact DA homogeneity $\sum$(all real number exponents) = whole number '1' for equation [or '3' for inequation]. *The proof for Condition 2 Theorem Riemann IV is now complete*□. Corollary 3.4 confirms de novo inexact DA homogeneity manifested as $\sum$(all real number exponents) = real number '$\neq 1$' for equation [or '$\neq 3$' for inequation] by all $\sigma \neq \frac{1}{2}$ values (non-critical lines) that are exclusively not associated with "actual location of all nontrivial zeros". Applying inclusion-exclusion principle: Exclusive presence of nontrivial zeros on critical line for Condition 2 Theorem Riemann IV implies exclusive absence of nontrivial zeros on non-critical lines for Condition 1 Theorem Riemann IV. *The proof for Condition 1 Theorem Riemann IV is now complete*□.

We logically deduce that explicit mathematical explanation why presence and absence of nontrivial zeros[6] should (respectively) coincide precisely with $\sigma = \frac{1}{2}$ and $\sigma \neq \frac{1}{2}$ [literally the Completely Predictable meta-properties ('overall' *complex properties*)] will require "complex" mathematical arguments. Attempting to provide explicit mathematical explanation with "simple" mathematical arguments would intuitively mean nontrivial zeros have to be (incorrectly and impossibly) treated as Completely Predictable entities.

Footnote 6: Completely Predictable meta-properties for Gram and virtual Gram points equating to "Presence of Gram[y=0] and Gram[x=0] points, and virtual Gram[y=0] and virtual Gram[x=0] points (respectively) coincide precisely with $\sigma = \frac{1}{2}$, and $\sigma \neq \frac{1}{2}$".

5. Prerequisite lemma, corollary and propositions for Gram[x=0] and Gram[y=0] conjectures

For Gram[y=0] and Gram[x=0] points (and corresponding virtual Gram[y=0] and virtual Gram[x=0] points with totally different values), we apply a parallel procedure carried out on nontrivial zeros but only depict abbreviated treatments and discussions.

Lemma 5.1. "Simplified" Gram[y=0] and Gram[x=0] points-Dirichlet eta functions are derived directly from Dirichlet eta function with Euler formula application and (respectively) they will intrinsically incorporate actual location [but not actual positions] of all Gram[y=0] and Gram[x=0] points.

Proof. For Gram[y=0] points, the equivalent of Eq. (4) and Eq. (6) are respectively given by Eq. (16) and Eq. (17) below.

$$\sum ReIm\{\eta(s)\} = Re\{\eta(s)\} + 0, \text{ or simply } Im\{\eta(s)\} = 0 \qquad (16)$$

$$\sum_{n=1}^{\infty}(2n)^{-\sigma}\sin(t\ln(2n)) = \sum_{n=1}^{\infty}(2n-1)^{-\sigma}\sin(t\ln(2n-1))$$

$$\sum_{n=1}^{\infty}(2n)^{-\sigma}\sin(t\ln(2n)) - \sum_{n=1}^{\infty}(2n-1)^{-\sigma}\sin(t\ln(2n-1)) = 0 \qquad (17)$$

For Gram[x=0] points, the equivalent of Eq. (4) and Eq. (6) are respectively given by Eq. (18) and Eq. (19) below.

$$\sum ReIm\{\eta(s)\} = 0 + Im\{\eta(s)\}, \text{ or simply } Re\{\eta(s)\} = 0 \qquad (18)$$

$$\sum_{n=1}^{\infty}(2n)^{-\sigma}\cos(t\ln(2n)) = \sum_{n=1}^{\infty}(2n-1)^{-\sigma}\cos(t\ln(2n-1))$$

$$\sum_{n=1}^{\infty}(2n)^{-\sigma}\cos(t\ln(2n)) - \sum_{n=1}^{\infty}(2n-1)^{-\sigma}\cos(t\ln(2n-1)) = 0 \qquad (19)$$

Eq. (17) and Eq. (19) being the "simplified" Gram[y=0] and Gram[x=0] points-Dirichlet eta functions derived directly from $\eta(s)$ will intrinsically incorporate *actual location [but not actual positions]* of (respectively) all Gram[y=0] and Gram[x=0] points. *The proof is now complete for Lemma 5.1*.□

Proposition 5.2. Gram[y=0] and Gram[x=0] points-Dirichlet Sigma-Power Laws in continuous (integral) format given as equations and inequations can both be (respectively) derived directly from "simplified" Gram[y=0] and Gram[x=0] points-Dirichlet eta functions in discrete (summation) format with Riemann integral application. [Note: Gram[y=0] and Gram[x=0] points-Dirichlet Sigma-Power Laws in continuous (integral) format here refers to relevant end-products obtained from "first key step of converting Riemann zeta function into its continuous format version".]

Proof. Antiderivatives below using (2n) parameter help obtain all subsequent equations: first two for Gram[y=0] points and second two for Gram[x=0] points.

$$\int_1^\infty (2n)^{-\sigma} \sin(t\ln(2n))dn =$$

$$\left[-\frac{(2n)^{1-\sigma}((\sigma-1)\sin(t\ln(2n)) + t\cos(t\ln(2n)))}{2(t^2+(\sigma-1)^2)} + C\right]_1^\infty$$

$$\int_1^\infty \sin(t\ln(2n))dn = \left[\frac{(2n)(\sin(t\ln(2n)) - t\cos(t\ln(2n)))}{2(t^2+1)} + C\right]_1^\infty$$

$$\int_1^\infty (2n)^{-\sigma} \cos(t\ln(2n))dn = \left[\frac{(2n)^{1-\sigma}(t\sin(t\ln(2n)) - (\sigma-1)\cos(t\ln(2n)))}{2(t^2+(\sigma-1)^2)} + C\right]_1^\infty$$

$$\int_1^\infty \cos(t\ln(2n))dn = \left[\frac{(2n)(t\sin(t\ln(2n)) + \cos(t\ln(2n)))}{2(t^2+1)} + C\right]_1^\infty$$

For Gram[y=0] points-Dirichlet Sigma-Power Law, the equivalent of Eq. (9) and Eq. (11) are respectively given by Eq. (20) as equation and Eq. (21) as inequation.

$$-\frac{1}{2(t^2+(\sigma-1)^2)} \cdot [(2n)^{1-\sigma}((\sigma-1)\sin(t\ln(2n)) + t\cos(t\ln(2n)))-$$

$$(2n-1)^{1-\sigma}((\sigma-1)\sin(t\ln(2n-1)) + t\cos(t\ln(2n-1)))]_1^\infty = 0 \qquad (20)$$

$$\left[\frac{(2n)(\sin(t\ln(2n)) - t\cos(t\ln(2n)))}{(2n-1)(\sin(t\ln(2n-1)) - t\cos(t\ln(2n-1)))} - \frac{(2n)^{\sigma+1}}{(2n-1)^{\sigma+1}}\right]_1^\infty \neq 0 \qquad (21)$$

For Gram[x=0] points-Dirichlet Sigma-Power Law, the equivalent of Eq. (9) and Eq. (11) are respectively given by Eq. (22) as equation and Eq. (23) as inequation.

$$\frac{1}{2(t^2+(\sigma-1)^2)} \cdot [(2n)^{1-\sigma}(t\sin(t\ln(2n)) - (\sigma-1)\cos(t\ln(2n)))-$$

$$(2n-1)^{1-\sigma}(t\sin(t\ln(2n-1)) - (\sigma-1)\cos(t\ln(2n-1)))]_1^\infty = 0 \qquad (22)$$

$$\left[\frac{(2n)(t\sin(t\ln(2n)) + \cos(t\ln(2n)))}{(2n-1)(t\sin(t\ln(2n-1)) + \cos(t\ln(2n-1)))} - \frac{(2n)^{\sigma+1}}{(2n-1)^{\sigma+1}}\right]_1^\infty \neq 0 \qquad (23)$$

Intended derivation of Gram[y=0] and Gram[x=0] points-Dirichlet Sigma-Power Laws as equations and inequations is successful. *The proof is now complete for Lemma 5.2*.□

Proposition 5.3. Exact Dimensional analysis homogeneity at $\sigma = \frac{1}{2}$ in Gram[y=0] and Gram[x=0] points-Dirichlet Sigma-Power Laws as equations and inequations are (respectively) indicated by $\sum$(all fractional exponents) = whole number '1' and '3'.

Proof. Gram[y=0] points-Dirichlet Sigma-Power Law as equation for $\sigma = \frac{1}{2}$ value is given by: $-\frac{1}{2t^2+\frac{1}{2}} \cdot [(2n)^{\frac{1}{2}}\left(t\cos(t\ln(2n)) - \frac{1}{2}\sin(t\ln(2n))\right) -$

$$(2n-1)^{\frac{1}{2}}\left(t\cos(t\ln(2n-1)) - \frac{1}{2}\sin(t\ln(2n-1))\right)\Big]_1^\infty = 0 \qquad (24)$$

Gram[y=0] points-Dirichlet Sigma-Power Law as inequation for $\sigma = \frac{1}{2}$ value is given by:

$$\left[\frac{(2n)(\sin(t\ln(2n)) - t\cos(t\ln(2n)))}{(2n-1)(\sin(t\ln(2n-1)) - t\cos(t\ln(2n-1)))} - \frac{(2n)^{\frac{3}{2}}}{(2n-1)^{\frac{3}{2}}}\right]_1^\infty \neq 0 \qquad (25)$$

Gram[x=0] points-Dirichlet Sigma-Power Law as equation for $\sigma = \frac{1}{2}$ value is given by: $\frac{1}{2t^2+\frac{1}{2}} \cdot [(2n)^{\frac{1}{2}}\left(t\sin(t\ln(2n)) + \frac{1}{2}\cos(t\ln(2n))\right) -$

$$\{(2n-1)^{\frac{1}{2}}\left(t\sin(t\ln(2n-1)) + \frac{1}{2}\cos(t\ln(2n-1))\right)\Big]_1^\infty = 0 \qquad (26)$$

Gram[x=0] points-Dirichlet Sigma-Power Law as inequation for $\sigma = \frac{1}{2}$ value is given by:

$$\left[\frac{(2n)(t\sin(t\ln(2n)) + \cos(t\ln(2n)))}{(2n-1)(t\sin(t\ln(2n-1)) + \cos(t\ln(2n-1)))} - \frac{(2n)^{\frac{3}{2}}}{(2n-1)^{\frac{3}{2}}}\right]_1^\infty \neq 0 \qquad (27)$$

$\sum$(all fractional exponents) as $2(1-\sigma)$ = whole number '1' for Eqs. (24) and (26), and $2(\sigma+1)$ = whole number '3' for Eqs. (25) and (27). These findings signify presence of complete sets Gram[y=0] points for Eqs. (24) and (25) and Gram[x=0] points for Eqs. (26) and (27). *The proof is now complete for Proposition 5.3*.□

Corollary 5.4. Inexact Dimensional analysis homogeneity at $\sigma \neq \frac{1}{2}$ [illustrated using $\sigma = \frac{2}{5}$] in Gram[y=0] and Gram[x=0] points-Dirichlet Sigma-Power Laws as equations and inequations are (respectively) indicated by $\sum$(all fractional exponents) = fractional number '$\neq 1$' and '$\neq 3$'.

Proof. Gram[y=0] points-Dirichlet Sigma-Power Law as equation for $\sigma = \frac{2}{5}$ value is given by: $-\frac{1}{2t^2+\frac{18}{25}} \cdot [(2n)^{\frac{3}{5}}\left(t\cos(t\ln(2n)) - \frac{3}{5}\sin(t\ln(2n))\right) -$

$$(2n-1)^{\frac{3}{5}}\left(t\cos(t\ln(2n-1)) - \frac{3}{5}\sin(t\ln(2n-1))\right)\Big]_1^\infty = 0 \qquad (28)$$

Gram[y=0] points-Dirichlet Sigma-Power Law as inequation for $\sigma = \frac{2}{5}$ value is given by:

$$\left[\frac{(2n)(\sin(t\ln(2n)) - t\cos(t\ln(2n)))}{(2n-1)(\sin(t\ln(2n-1)) - t\cos(t\ln(2n-1)))} - \frac{(2n)^{\frac{7}{5}}}{(2n-1)^{\frac{7}{5}}}\right]_1^{\infty} \neq 0 \qquad (29)$$

Gram[x=0] points-Dirichlet Sigma-Power Law as equation for $\sigma = \frac{2}{5}$ value is given by:
$\frac{1}{2t^2 + \frac{18}{25}} \cdot [(2n)^{\frac{3}{5}} \left(t\sin(t\ln(2n)) + \frac{3}{5}\cos(t\ln(2n))\right) -$

$$(2n-1)^{\frac{3}{5}} \left(t\sin(t\ln(2n-1)) + \frac{3}{5}\cos(t\ln(2n-1))\right)]_1^{\infty} = 0 \qquad (30)$$

Gram[x=0] points-Dirichlet Sigma-Power Law as inequation for $\sigma = \frac{2}{5}$ value is given by:

$$\left[\frac{(2n)(t\sin(t\ln(2n)) + \cos(t\ln(2n)))}{(2n-1)(t\sin(t\ln(2n-1)) + \cos(t\ln(2n-1)))} - \frac{(2n)^{\frac{7}{5}}}{(2n-1)^{\frac{7}{5}}}\right]_1^{\infty} \neq 0 \qquad (31)$$

$\sum$(all fractional exponents) as $2(1-\sigma)$ = fractional number '$\neq 1$' for Eqs. (28) and (30), and $2(\sigma+1)$ = fractional number '$\neq 3$' for Eqs. (29) and (31). These findings signify presence of complete sets virtual Gram[y=0] points for Eqs. (28) and (29) and virtual Gram[x=0] points for Eqs. (30) and (31). *The proof is now complete for Corollary 5.4*□.

6. Prime and Composite numbers

Prime and Composite numbers are Incompletely Predictable entities dependently linked together in a sequential, cummulative and eternal manner because the relationship Number '1' + Prime numbers + Composite numbers = Natural numbers is always valid for all Natural numbers.

6.1 Dimensional analysis on Cardinality and "Dimensions" for Prime numbers

We use the word "Dimensions" to denote well-defined Incompletely Predictable entities obtained from Information-Complexity conservation. Relevant "Dimensions" *dependently* represent Number '1', **P** and **C**. Then *by default* any (sub)sets of **P** and **C** in well-defined equations can also be represented by their corresponding "Dimensions".

Remark 6.1. We can apply Dimensional analysis to "Dimensions" from Information-Complexity conservation and cardinality of relevant sets in certain well-defined equations.

Let **X** denote **E**, **O**, **N** [which are classified as Completely Predictable numbers], **P** and **C** [which are classified as Incompletely Predictable numbers]. For x = 1, 2, 3, 4, 5,..., ∞; consider all **X** $\leqslant$ x. Then this "all **X** $\leqslant$ x" is definition for **X**-$\pi(x)$ [denoting "**X** counting function"] resulting in following two types of equations coined as (I) 'Exact' equation **N**-$\pi(x)$ = **E**-$\pi(x)$ + **O**-$\pi(x)$ with "non-varying" relationships **E**-$\pi(x)$ = **O**-$\pi(x)$ for all x = **E** and **E**-$\pi(x)$ = **O**-$\pi(x)$ - 1 for all x = **O**, and (II) 'Inexact' equation **N**-$\pi(x)$ = 1 + **P**-$\pi(x)$ + **C**-$\pi(x)$ with "varying" relationships **P**-$\pi(x)$ > **C**-$\pi(x)$ for all x $\leqslant$ 8; **P**-$\pi(x)$ = **C**-$\pi(x)$ for x = 9, 11, and 13; and **P**-$\pi(x)$ < **C**-$\pi(x)$ for x = 10, 12, and all x $\geqslant$ 14.

Let "Dimensions" and different (sub)sets of **E**, **O**, **N**, **P** and **C** be 'base quantities'. Then exponent '1' of "Dimensions" and cardinality of these (sub)sets in well-defined equations are corresponding 'units of measurement'. Performing DA on "Dimensions" for **PC** pairing is depicted later on. Performing DA on cardinality is depicted next.

For Set **N** = Set **E** + Set **O**, then $|\mathbf{N}| = |\mathbf{E}| + |\mathbf{O}| \implies \aleph_0 = \aleph_0 + \aleph_0$ thus conforming with DA homogeneity.

For Set **N** = Set **P** + Set **C** + Number '1', then Set **N** - Number '1' = Set **P** + Set **C** and $|\mathbf{N}$ - Number '1'$| = |\mathbf{P}| + |\mathbf{C}| \implies \aleph_0 = \aleph_0 + \aleph_0$ thus conforming with DA homogeneity.

For Set **N** - Set **even P** - Number '1'= Set **odd P** + Set **even C** + Set **odd C**, then $|\mathbf{N}$ - **even P** - Number '1'$| = |\mathbf{odd\ P}| + |\mathbf{even\ C}| + |\mathbf{odd\ C}| \implies \aleph_0 = \aleph_0 + \aleph_0 + \aleph_0$ thus conforming with DA homogeneity. Symbolically represented by all available **O** prime gap = 1 and **E** prime gaps = 2, 4, 6, 8, 10,...; **O** composite gap = 1 and **E** composite gap = 2; and **O** natural gap = 1; then $|\mathbf{Gap\ 1\ N}$ - **Gap 1 P** - Number '1'$| = |\mathbf{Gap\ 2\ P}| + |\mathbf{Gap\ 4\ P}| + |\mathbf{Gap\ 6\ P}| + |\mathbf{Gap\ 8\ P}| + |\mathbf{Gap\ 10\ P}| + ... + |\mathbf{Gap\ 1\ C}| + |\mathbf{Gap\ 2\ C}| \implies \aleph_0 = \aleph_0 + \aleph_0 + \aleph_0 + \aleph_0 + \aleph_0 + ... \aleph_0 + \aleph_0$ thus conforming with DA homogeneity. It is known that $|\mathbf{Gap\ 1\ P}| = |$Number '1'$| = 1$ and $|\mathbf{Gap\ 1\ N}| = |\mathbf{Gap\ 1\ C}| = |\mathbf{Gap\ 2\ C}| = \aleph_0$. Then solving Polignac's and Twin prime conjectures translate to successfully proving $|\mathbf{Gap\ 2\ P}| = |\mathbf{Gap\ 4\ P}| = |\mathbf{Gap\ 6\ P}| = |\mathbf{Gap\ 8\ P}| = |\mathbf{Gap\ 10\ P}| = ... = \aleph_0$ with $|\mathbf{E\ prime\ gaps}| = \aleph_0$.

Outline of proof for Polignac's and Twin prime conjectures. Requires simultaneously satisfying two mutually inclusive conditions: I. *With rigid manifestation of DA homogeneity*, quantitive[7] fulfillment by considering i ∈ **E** for each Subset **odd** $\mathbf{P}_i$ generated by **E** prime gap = i from Set **E prime gaps** occurs only if solitary cardinality value is present in equation Set **odd P** = $\sum_{i=2}^{\infty}$ Subset **odd** $\mathbf{P}_i$ with $|\mathbf{odd\ P}| = |\mathbf{odd\ P}_i| = |\mathbf{E\ prime\ gaps}| = \aleph_0$, and II. *With rigid manifestation of DA non-homogeneity*, quantitive[7] fulfillment by considering i ∈ **E** for each Subset **odd** $\mathbf{P}_i$ generated by **E** prime gap = i from Set **E prime gaps** does not occur if more than one cardinality values are present in equation Set **odd P** > $\sum_{i=2}^{\infty}$ Subset **odd** $\mathbf{P}_i$ with $|\mathbf{E\ prime\ gaps}| = \aleph_0$ having incorrect |Subset(s) **odd P**| = N (finite value) and/or Set **odd P** > $\sum_{i=2}^{N}$ Subset **odd** $\mathbf{P}_i$ with $|\mathbf{odd\ P}_i| = \aleph_0$ having incorrect $|\mathbf{E\ prime\ gaps}| = N$ (finite value).

Footnote 7: Qualitative fulfillment of $|\mathbf{odd\ P}| = |\mathbf{odd\ P}_i| = |$all **E prime gaps**$| = \aleph_0$ equates to Plus-Minus Gap 2 Composite Number Alternating Law being precisely obeyed by all **E** prime gaps apart from first **E** prime gap precisely obeying Plus Gap 2 Composite Number Continuous Law. Derived using Information-Complexity conservation, these Laws symbolize "end-result" proof on Polignac's and Twin prime conjectures. *Law of Continuity* is a heuristic principle *whatever succeed for the finite, also succeed for the infinite*. Then these Laws which inherently manifest 'Gap 2 Composite Number' on finite and infinite time scale should in principle "succeed for the finite, also succeed for the infinite".

Polignac's and Twin prime conjectures mathematical foot-prints. Six identifiable steps to prove these conjectures: *Step 1* Considering x ∈ **N**, obtain Dimensions $(2x - 2)^1$, $(2x - 4)^1$, $(2x - 5)^1$, $(2x - 7)^1$, $(2x - 8)^1$, $(2x - 9)^1$, ..., $(2x - \infty)^1$ with specific groupings to constitute all elements of Set **P** [culminating in obtaining all prime gaps (= **E** prime gaps + Solitary **O** prime gap) with $|\mathbf{all\ prime\ gaps}| = \aleph_0$]. Note Dimension $(2x - 2)^1$ represents x = 1 (Number '1') which is neither **P** nor **C**. *Step 2* Considering i ∈ **E**, confirm perpetual recurrences of individual **E** prime gap = i (associated with its unique odd $\mathbf{P}_i$) occur only when depicted as specific groupings of these Dimensions endowed with exponent '1' for all ranges of x. *Step 3* Perform DA on exponent

'1' in these Dimensions. *Step 4* Perform DA on equation Set **odd P** = $\sum_{i=2}^{\infty}$ Subset **odd** $\mathbf{P}_i$ to obtain |**odd P**| = |**odd** $\mathbf{P}_i$| = $\aleph_0$ whereby Subset **odd** $\mathbf{P}_i$ is derived from its associated unique **E** prime gap = i with |**E prime gaps**| = $\aleph_0$. *Step 5* Confirm 'Prime number' variable and 'Prime gap' variable complex algorithm "containing" all **P** with knowing their overall actual location [but not actual positions][8]. *Step 6* Derive Plus-Minus Gap 2 Composite Number Alternating Law and Plus Gap 2 Composite Number Continuous Law using Information-Complexity conservation.

Footnote 8: This phrase implies all **P** (and **C**) are treated as Incompletely Predictable numbers. Actual positions will require using complex algorithm Sieve of Eratosthenes to *dependently* calculate positions of all preceding **P** (and **C**) in the neighborhood.

'Complex Elementary Fundamental Laws'-based solutions of Plus-Minus Gap 2 Composite Number Alternating Law and Plus Gap 2 Composite Number Continuous Law are obtained by undertaking the non-negotiable mathematical steps outlined above. These Laws are literally Completely Predictable meta-properties ('overall' *complex properties*) arising from "interactions" between **P** and **C** producing relevant patterns of Gap 2 Composite Number perpetual appearances [albeit with Incompletely Predictable timing]. We logically deduce explicit mathematical explanation for these meta-properties requires "complex" mathematical arguments. Attempts to give explicit mathematical explanation with "simple" mathematical arguments would intuitively mean Incompletely Predictable numbers **P** and **C** be (incorrectly and impossibly) treated as Completely Predictable numbers.

6.2 Brief overview of Polignac's and Twin prime conjectures

Occurring over 2000 years ago (c. 300 BC), ancient Euclid's proof on infinitude of **P** in totality [viz. |**P**| = $\aleph_0$ for Set **P**] predominantly by *reductio ad absurdum* (proof by contradiction) is earliest known but not the only proof for this simple problem in Number theory. Since then dozens of proofs have been devised such as three chronologically listed: Goldbach's Proof using Fermat numbers (written in a letter to Swiss mathematician Leonhard Euler, July 1730), Furstenberg's Topological Proof in 1955[5], and Filip Saidak's Proof in 2006[6]. The strangest candidate is likely to be Furstenberg's Topological Proof.

In 2013, Yitang Zhang proved a landmark result showing some unknown even number 'N' < 70 million such that there are infinitely many pairs of **P** that differ by 'N'[7]. By optimizing Zhang's bound, subsequent Polymath Project collaborative efforts using a new refinement of GPY sieve in 2013 lowered 'N' to 246; and assuming Elliott-Halberstam conjecture and its generalized form have further lower 'N' to 12 and 6, respectively. Then 'N' has intuitively more than one valid values such that there are infinitely many pairs of **P** that differ by each of those 'N' values [thus proving existence of more than one Subset **odd** $\mathbf{P}_i$ with |**odd** $\mathbf{P}_i$| = $\aleph_0$]. We can only theoretically lower 'N' to 2 (in regards to **P** with 'small gaps') but there are still an infinite number of **E** prime gaps (in regards to **P** with 'large gaps') that require "the proof that each will generate its unique set of infinite **P**".

Remark 6.2. Existence of maximal and non-maximal prime gaps supply crucial indirect evidence to intuitively support but does not prove "Each even prime gap will generate an infinite magnitude of odd prime numbers on its own accord".

Comments relevant to Remark 6.2 are given in Section 7 below.

Creed Odyssey in Mathematics and Medicine series

TABLE 1. First 17 prime gaps depicted in the format utilizing maximal prime gaps [depicted with asterisk symbol (*)] and non-maximal prime gaps [depicted without this asterisk symbol].

Prime gap	Following prime number	Prime gap	Following prime number
1*	2	18*	523
2*	3	20*	887
4*	7	22*	1129
6*	23	24	1669
8*	89	26	2477
10	139	28	2971
12	199	30	4297
14*	113	32	5591
16	1831	..	....

7. Supportive role of maximal and non-maximal prime gaps

We analyze data of all **P** obtained when extrapolated out over a wide range of $x \geq 2$ integer values. As sequence of **P** carries on, **P** with ever larger prime gaps will appear. For given range of x integer values, prime gap = n_2 is a 'maximal prime gap' if prime gap = n_1 < prime gap = n_2 for all $n_1 < n_2$. In other words, the largest such prime gaps in this range are called maximal prime gaps. The term 'first occurrence prime gaps' refers to first occurrences of maximal prime gaps whereby maximal prime gaps are prime gaps of "at least of this length".

We use maximal prime gaps to denote 'first occurrence prime gaps'. CIS non-maximal prime gaps (endorsed with nickname 'slow jumpers') will always lag behind CIS maximal prime gaps for onset appearances in **P** sequence. These are shown for first 17 prime gaps in Table 1. Apart from **O** prime gap = 1 representing solitary even **P** '2', remaining **P** in Table 1 consist of representative single odd **P** for each **E** prime gap. These odd **P** individually make one-off appearance in **P** sequence in a *perpetual albeit Incompletely Predictable manner*. Initial seven of [majority] "missing" odd **P** are 5, 11, 13, 17, 19, 29, 31,... belonging to Subset **P** with 'residual' prime gaps are potential source of odd **P** in relation to proposal that each **E** prime gap from Set **E** **prime gaps** will generate its specific Subset **odd P**. Set all **P** from all prime gaps = Subset **P** from maximal prime gaps + Subset **P** from non-maximal prime gaps + Subset **P** from 'residual' prime gaps. Subset **P** from 'residual' prime gaps with representation from all **E** prime gaps must include all correctly selected "missing" odd **P**. These observations support but does not prove proposition that each **E** prime gap will generate its own Subset **odd P** with $|\text{odd } \mathbf{P}| = \aleph_0$.

For $i \in \mathbf{N}$; primordial $P_i\#$ is analog of usual factorial for **P** = 2, 3, 5, 7, 11, 13,.... Then $P_1\#$ = 2, $P_2\# = 2 \times 3 = 6$, $P_3\# = 2 \times 3 \times 5 = 30$, $P_4\# = 2 \times 3 \times 5 \times 7 = 210$, $P_5\# = 2 \times 3 \times 5 \times 7 \times 11 = 2310$, $P_6\# = 2 \times 3 \times 5 \times 7 \times 11 \times 13 = 30030$, etc. English mathematician John Horton Conway coined the term 'jumping champion' in 1993. An integer n is a 'jumping champion' if n is the most frequently occurring difference (prime gap) between consecutive **P**<x for some x integer values. Example: for any x with 7<x<131, n = 2 (indicating twin **P**) is the 'jumping champion'. It has been conjectured that (i) the only 'jumping champions' are 1, 4 and primorials 2, 6, 30, 210, 2310, 30030,... and (ii) 'jumping champions' tend to infinity. Their required proofs will likely need proof of k-tuple conjecture. **P** from 'jumping champion' prime gaps have their onset appearances in **P** sequence in a *perpetual albeit Incompletely Predictable manner* [as another example to that outlined in previous paragraph].

Book 2 Three Open Problems by Riemann and Polignac

8. Information-Complexity conservation

A formula, as equation or algorithm, is simply a Black Box generating necessary Output (with qualitative structural 'Complexity') when supplied with given Input (with quantitative data 'Information'). This 'Information' and 'Complexity' are what is referred to in the term 'Information-Complexity conservation'.

N (CIS): 1, 2, 3,..., $+\infty$. Let x be from Set **X** such that x $\in$ **N**. Consider x for upper boundary of interest in Set **X** whereby **X** is chosen from **N**, **E**, **O**, **P** or **C**.

Lemma 8.1. Natural counting function **N**-$\pi(x)$, defined as $|\mathbf{N} \leqslant x|$, is Completely Predictable by independently using simple algorithm to be equal to x.

Proof Formula to generate **N** with 100% certainty is $\mathbf{N}_i = i$ whereby $\mathbf{N}_i$ is the ith **N** and i = 1, 2, 3,..., ∞. For a given $\mathbf{N}_i$, its ith position is simply i. Natural gap $(\mathbf{G}_{Ni}) = \mathbf{N}_{i+1} - \mathbf{N}_i$, with $\mathbf{G}_{Ni}$ always = 1. There are x **N** $\leqslant$ x. Thus **N**-$\pi(x) = |\mathbf{N} \leqslant x| = x$. *The proof is now complete for Lemma 8.1*□.

Lemma 8.2. Even counting function **E**-$\pi(x)$, defined as $|\mathbf{E} \leqslant x|$, is Completely Predictable by independently using simple algorithm to be equal to floor(x/2).

Proof. Formula to generate **E** with 100% certainty is $\mathbf{E}_i = i\text{X}2$ whereby $\mathbf{E}_i$ is the ith **E** and i = 1, 2, 3,..., ∞ abiding to mathematical label "All **N** always ending with a digit 0, 2, 4, 6 or 8". For a given $\mathbf{E}_i$, its ith position is calculated as i = $\mathbf{E}_i/2$. Even gap $(\mathbf{G}_{Ei}) = \mathbf{E}_{i+1} - \mathbf{E}_i$, with $\mathbf{G}_{Ei}$ always = 2. There are $\lfloor \frac{x}{2} \rfloor$ **E** $\leqslant$ x. Thus **E**-$\pi(x) = |\mathbf{E} \leqslant x| = $ floor(x/2). *The proof is now complete for Lemma 8.2*□.

Lemma 8.3. Odd counting function **O**-$\pi(x)$, defined as $|\mathbf{O} \leqslant x|$, is Completely Predictable by independently using simple algorithm to be equal to ceiling(x/2).

Proof. Formula to generate **O** with 100% certainty is $\mathbf{O}_i = (i\text{X}2) - 1$ whereby $\mathbf{O}_i$ is the ith odd number and i = 1, 2, 3,..., ∞ abiding to mathematical label "All **N** always ending with a digit 1, 3, 5, 7, or 9". For a given $\mathbf{O}_i$ number, its ith position is calculated as i = $(\mathbf{O}_i + 1)/2$. Odd gap $(\mathbf{G}_{Oi}) = \mathbf{O}_{i+1} - \mathbf{O}_i$, with $\mathbf{G}_{Oi}$ always = 2. There are $\lceil \frac{x}{2} \rceil$ **O** $\leqslant$ x. Thus **O**-$\pi(x) = |\mathbf{O} \leqslant x| = $ ceiling(x/2). *The proof is now complete for Lemma 8.3*□.

Lemma 8.4. Prime counting function **P**-$\pi(x)$, defined as $|\mathbf{P} \leqslant x|$, is Incompletely Predictable with Set **P** dependently obtained using complex algorithm Sieve of Eratosthenes.

Proof. Algorithm to generate $\mathbf{P}_i$ whereby $\mathbf{P}_1$ (= 2), $\mathbf{P}_2$ (= 3), $\mathbf{P}_3$ (= 5), $\mathbf{P}_4$ (= 7),..., ∞ with 100% certainty is based on Sieve of Eratosthenes abiding to mathematical label "All **N** apart from 1 that are evenly divisible by itself and by 1". Although we can check primality of a given **O** by trial division, we can never determine its position without knowing positions of preceding **P**. Prime gap $(\mathbf{G}_{Pi}) = \mathbf{P}_{i+1} - \mathbf{P}_i$, with $\mathbf{G}_{Pi}$ constituted by all **E** except 1st $\mathbf{G}_{P1} = 3 - 2 = 1$. **P**-$\pi(x) = |\mathbf{P} \leqslant x|$. This is Incompletely Predictable and is calculated via mentioned algorithm. Using definition of prime gap, every **P** [represented here with aid of 'n' notation instead of usual 'i' notation] is written as $\mathbf{P}_{n+1} = 2 + \sum_{i=1}^{n} \mathbf{G}_{Pi}$ with '2' denoting $\mathbf{P}_1$. Here i & n = 1, 2, 3, 4, 5, ..., ∞. *The proof is now complete for Lemma 8.4*□.

Lemma 8.5. Composite counting function **C**-$\pi(x)$, defined as $|\mathbf{C} \leqslant x|$, is Incompletely Predictable with Set **C** derived as Set **N** - Set **P** [dependently obtained using complex algorithm Sieve of Eratosthenes] - Number '1'.

Proof. Composite numbers abide to mathematical label "All **N** apart from 1 that are evenly divisible by numbers other than itself and 1". Algorithm to generate $\mathbf{C}_i$ whereby $\mathbf{C}_1$ (= 4), $\mathbf{C}_2$ (=

6), C_3 (= 8), C_4 (= 9),..., ∞ with 100% certainty is based [indirectly] on Sieve of Eratosthenes via selecting non-prime **N** to be **C**. We define Composite gap G_{Ci} as C_{i+1} - C_i with G_{Ci} constituted by 1 & 2. **C**-$\pi(x)$ = **C** $\leqslant$ x. This is Incompletely Predictable and always need to be calculated indirectly via mentioned algorithm. Using definition of composite gap, every **C** [represented here with aid of 'n' notation instead usual 'i' notation] is written as $C_{n+1} = 4 + \sum_{i=1}^{n} G_{Ci}$ with '4' denoting C_1. Here i & n = 1, 2, 3, 4, 5, ..., ∞. *The proof is now complete for Lemma 8.5*□.

Denote **X** to be **N**, **E**, **O**, **P** or **C**. **X**-$\pi(x)$ = |**X** $\leqslant$ x| with x $\in$ **N**. We define and compute entity 'Grand-Total Gaps for **X** at x' (Grand-Total $\Sigma \mathbf{X}_x$-Gaps).

Proposition 8.6. For any given x $\geqslant$ 1 values in Set **N**, designated Complexity is represented by $\Sigma \mathbf{N}_x$-Gaps = x - N with N = 1 being maximal.

Proof. Set **N** (for x = 1 to 12): 1, 2, 3, 4, 5, 6, 7, 8, 9, 10, 11, 12. **N**-$\pi(x)$ = 12. There are x - 1 = 11 **N**-Gaps each of '1' magnitude: 1, 1, 1, 1, 1, 1, 1, 1, 1, 1, 1. $\Sigma \mathbf{N}_x$-Gaps = 11 X 1 = 11. This equates to "x - 1" – regarded as Complexity for **N**. *The proof is now complete for Proposition 8.6*□.

Proposition 8.7. For any given x $\geqslant$ 1 values in constituent Set **E** and Set **O**, designated Complexity is represented by $\Sigma \mathbf{EO}_x$-Gaps = 2x - N with N = 4 being maximal.

Proof. Set **E** and Set **O** (for x = 1 to 12): 2, 4, 6, 8, 10, 12 and 1, 3, 5, 7, 9, 11. **E**-$\pi(x)$ = 6 and **O**-$\pi(x)$ = 6. There are $\lfloor \frac{x}{2} \rfloor$ - 1 = 5 **E**-Gaps each of '2' magnitude: 2, 2, 2, 2, 2. $\Sigma \mathbf{E}_x$-Gaps = 5 X 2 = 10, and $\lceil \frac{x}{2} \rceil$ - 1 = 5 **O**-Gaps each of '2' magnitude: 2, 2, 2, 2, 2. $\Sigma \mathbf{O}_x$-Gaps = 5 X 2 = 10. Grand-Total $\Sigma \mathbf{EO}_x$-Gaps = 10 + 10 = 20. Depicted by Table 3 and Figure 9 in Appendix I, 2x - N = "2x - 4" [perpetual constant appearances of "N = 4 being maximal"] is Complexity for **E** and **O**. *The proof is now complete for Proposition 8.7*□.

Proposition 8.8. For selected x $\geqslant$ 2 values in constituent Set **P** and Set **C**, designated Complexity is cyclically represented by $\Sigma \mathbf{PC}_x$-Gaps = 2x - N with N = 7 being minimal.

Proof. Set **P** and Set **C** (for x = 2 to 12): 2, 3, 5, 7, 11 and 4, 6, 8, 9, 10, 12. **P**-$\pi(x)$ = 5 and **C**-$\pi(x)$ = 6. There are four **P**-Gaps of 1, 2, 2, 4 magnitude and five **C**-Gaps of 2, 2, 1, 1, 2 magnitude. $\Sigma \mathbf{P}_x$-Gaps = 1 + 2 + 2 + 4 = 9. $\Sigma \mathbf{C}_x$-Gaps = 2 + 2 + 1 + 1 + 2 = 8. Grand-Total $\Sigma \mathbf{PC}_x$-Gaps = 9 + 8 = 17. Depicted by Table 2 and Figure 8, 2x - N = "2x - 7" [perpetual intermittent and cyclical appearances of "N = 7 being minimal"] is Complexity for **P** and **C**. *The proof is now complete for Proposition 8.8*□.

Designated Complexity is (i) x - N with N = 1 (maximal) for Completely Predictable **N**, (ii) 2x - N with N = 7 (minimal) for Incompletely Predictable **P** & **C**, and (iii) 2x - N with N = 4 (maximal) for Completely Predictable **E** & **O**. Interpretations: **N** has minimal Complexity, **E** & **O** have intermediate Complexity, and **P** & **C** have maximal [varying] Complexity. Defacto baseline "2x - 4" Grand-Total Gaps [minus 4 value] in **E**-**O** pairing > Defacto baseline "2x - $\geqslant$7" Grand-Total Gaps [minus $\geqslant$7 values] in **P**-**C** pairing.

Let both x & N $\in$ **N**. We tabulate in Table 2 and graph in Figure 8 [Incompletely Predictable] **P**-**C** mathematical landscape for a relatively larger x = 2 to 64 here (and ditto for [Completely Predictable] **E**-**O** mathematical landscape for relatively larger x = 1 to 64 in Appendix D). The term "mathematical landscape" denotes specific mathematical patterns in tabulated and graphed data. "Dimension" contextually denotes Dimension 2x - N whereby (i) allocated [infinite] N values result in Dimensions 2x - 7, 2x - 8, 2x - 9, ..., 2x - ∞ for **P**-**C** finite scale mathematical landscape and (ii) allocated [finite] N values for **E**-**O** finite scale mathematical landscape result in Dimension 2x - 4. For **P**-**C** pairing, initial one-off Dimensions 2x - 2, 2x - 4 and 2x - 5 (in consecutive order) are exceptions [with Dimension 2x - 2 validly representing Number '1' which is neither **P** nor **C**]. For **E-O** pairing, initial one-off Dimension 2x - 2 is an exception. **P-C** mathematical landscape consisting of Dimensions will intrinsically incorporate **P** and **C** in an integrated manner and there are infinite times whereby relevant Dimensions deviate away from 'baseline' Dimension 2x - 7 simply because **P** [and, by default, **C**] in totality are rigorously proven to be infinite in magnitude. In contrast, there is a complete lack of deviation away from 'baseline' Dimension 2x - 4 apart from one-off deviation caused by the initial Dimension 2x - 2 in Appendix D.

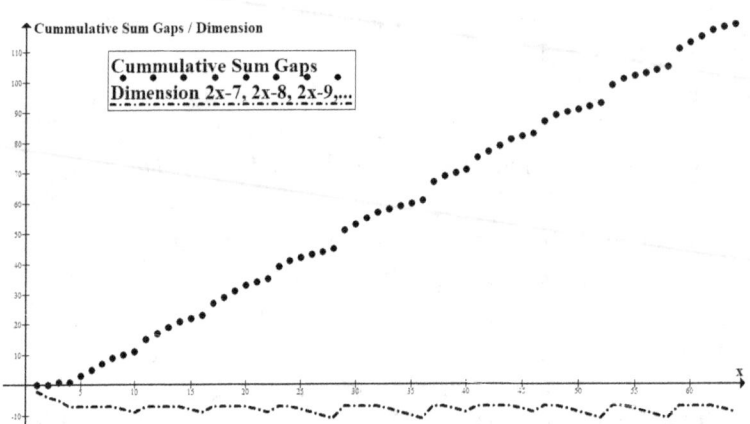

FIGURE 8. Prime-Composite finite scale mathematical (graphed) landscape using data obtained for x = 2 to 64. Bottom graph symbolically represent "Dimensions" using ever larger negative integers.

In Figure 8, Dimensions 2x - 7, 2x - 8, 2x - 9, ..., 2x - ∞ are symbolically represented by -7, -8, -9, ..., ∞ with 2x - 7 displayed as 'baseline' Dimension whereby Dimension trend (Cumulative Sum Gaps) must repeatedly reset itself onto this 'baseline' Dimension on a perpetual basis. Dimensions symbolically represented by ever larger negative integers will correspond to **P** associated with ever larger prime gaps and this phenomenon will generally happen at ever larger x values (with complete presence of Chaos and Fractals being manifested in our graph). At ever larger x values, **P**-$\pi(x)$ will overall become larger but with a *decelerating* trend whereas **C**-$\pi(x)$ will overall become larger but with an *accelerating* trend. This support ever larger prime gaps appearing at ever larger x values. Definitive derivation of data in Table 2 is illustrated by two examples for position x = 31 & 32. For i & x ∈ **N**; ΣPC$_x$-Gap = ΣPC$_{x-1}$-Gap + Gap value at **P**$_{i-1}$ or Gap value at **C**$_{i-1}$ whereby (i) **P**$_i$ or **C**$_i$ at position x is determined by whether relevant x value belongs to a **P** or **C**, and (ii) both ΣPC$_1$-Gap and ΣPC$_2$-Gap = 0. Example, for position x = 31: 31 is **P** (**P11**). Desired Gap value at **P10** = 2. Thus ΣPC$_{31}$-Gap (55) = ΣPC$_{30}$-Gap (53) + Gap value at **P10** (2). Example, for position x = 32: 32 is **C** (**C20**). Desired Gap value at **C19** = 2. Thus ΣPC$_{32}$-Gap (57) = ΣPC$_{31}$-Gap (55) + Gap value at **C20** (2). 'Overall magnitude of **C** will always be greater than that of **P**' will hold true from x = 14 onwards. For instance, position x = 61 corresponds to **P** 61 which is 18th **P**, whereas [the one lower] position x = 60 corresponding to **C** 60 is the [much higher] 42nd **C**.

TABLE 2. Prime-Composite finite scale mathematical (tabulated) landscape using data obtained for x = 2 to 64. The Number '1' is neither prime nor composite. Legend: C = composite, P = prime, Dim = Dimension, Y = Dimension 2x - 7 (for visual clarity), N/A = Not Applicable.

x	P_i or C_i, Gaps	ΣPC_x-Gaps	Dim	x	P_i or C_i, Gaps	ΣPC_x-Gaps	Dim
1	N/A	0	2x-2	33	C21, 1	58	2x-8
2	P1, 1	0	2x-4	34	C22, 1	59	2x-9
3	P2, 2	1	2x-5	35	C23, 1	60	2x-10
4	C1, 2	1	Y	36	C24, 2	61	2x-11
5	P3, 2	3	Y	37	P12, 4	67	Y
6	C2, 2	5	Y	38	C25, 1	69	Y
7	P4, 4	7	Y	39	C26, 1	70	2x-8
8	C3, 1	9	Y	40	C27, 1	71	2x-9
9	C4, 1	10	2x-8	41	P13, 2	75	Y
10	C5, 2	11	2x-9	42	C28, 2	77	Y
11	P5, 2	15	Y	43	P14, 4	79	Y
12	C6, 2	17	Y	44	C29, 1	81	Y
13	P6, 4	19	Y	45	C30, 1	82	2x-8
14	C7, 1	21	Y	46	C31, 2	83	2x-9
15	C8, 1	22	2x-8	47	P15, 6	87	Y
16	C9, 1	23	2x-9	48	C32, 1	89	Y
17	P7, 2	27	Y	49	C33, 1	90	2x-8
18	C10, 2	29	Y	50	C34, 1	91	2x-9
19	P8, 4	31	Y	51	C35, 1	92	2x-10
20	C11, 1	33	Y	52	C36, 1	93	2x-11
21	C12, 1	34	2x-8	53	P16, 6	99	Y
22	C13, 2	35	2x-9	54	C37, 1	101	Y
23	P9, 6	39	Y	55	C38, 1	102	2x-8
24	C14, 1	41	Y	56	C39, 1	103	2x-9
25	C15, 1	42	2x-8	57	C40, 1	104	2x-10
26	C16, 1	43	2x-9	58	C41, 1	105	2x-11
27	C17, 1	44	2x-10	59	P17, 2	111	Y
28	C18, 2	45	2x-11	60	C42, 2	113	Y
29	P10, 2	51	Y	61	P18, 6	115	Y
30	C19, 2	53	Y	62	C43, 1	117	Y
31	P11, 6	55	Y	63	C44, 1	118	2x-8
32	C20, 1	57	Y	64	C45, 1	119	2x-9

Book 2 Three Open Problems by Riemann and Polignac

9. Polignac's and Twin prime conjectures

Previous section alludes to **P-C** finite scale mathematical landscape. This section alludes to **P-C** infinite scale mathematical landscape. Let 'Y' symbolizes (baseline) Dimension 2x - 7. Let prime gap at $\mathbf{P}_i = \mathbf{P}_{i+1} - \mathbf{P}_i$ with $\mathbf{P}_i$ & $\mathbf{P}_{i+1}$ respectively symbolizes consecutive "first" & "second" **P** in any $\mathbf{P}_i$-$\mathbf{P}_{i+1}$ pairings. We denote (i) Dimensions YY grouping [depicted by 2x - 7 initially appearing twice in (iii)] to represent signal for appearances of **P** pairings other than twin **P** such as cousin **P**, sexy **P**, etc; (ii) Dimension YYYY grouping to represent signal for appearances of **P** pairings as twin **P**; and (iii) Dimension (2x - $\geqslant$7)-Progressive-Grouping allocated to 2x - 7, 2x - 7, 2x - 8, 2x - 9, 2x - 10, 2x - 11,..., 2x - ∞ as elements of *precise* and *proportionate* CFS Dimensions representation of an individual $\mathbf{P}_i$ with its associated prime gap namely, Dimensions 2x - 7 & 2x - 7 pairing = twin **P** (with both its prime gap & CFS cardinality = 2); 2x - 7, 2x - 7, 2x - 8 & 2x - 9 pairing = cousin **P** (with both its prime gap & CFS cardinality = 4); 2x - 7, 2x - 7, 2x - 8, 2x - 9, 2x - 10 & 2x - 11 pairing = sexy **P** (with both its prime gap & CFS cardinality = 6); and so on. The higher order [traditionally defined as closest possible] prime groupings of three **P** as **P** triplets, of four **P** numbers as prime quadruplets, of five **P** numbers as prime quintuplets, etc consist of relevant serendipitous groupings abiding to mathematical rule: With exception of three 'outlier' **P** 3, 5, & 7; groupings of any three **P** as **P**, **P**+2, **P**+4 combination (viz. manifesting two consecutive twin **P**) is a mathematical impossibility. The 'anomaly' one of every three consecutive **O** is a multiple of three, and hence this particular number cannot be **P**, explains this impossibility. Then closest possible **P** grouping [viz. for prime triplet] must be either **P**, **P**+2, **P**+6 format or **P**, **P**+4, **P**+6 format.

P groupings not respecting traditional closest-possible-prime groupings are also the norm occurring infinitely often, indicating continual presence of prime gaps $\geqslant$ 6. As **P** become sparser at larger range, perpetual presence of (i) prime gaps $\geqslant$ 6 [which we propose to arbitrarily represent 'large gaps'] and (ii) prime gaps 2 & 4 [which we propose to arbitrarily represent 'small gaps'] with progressive greater magnitude will cummulatively occur for each prime gap but always in a decelerating manner. With permanent requirement at larger range of intermittently resetting to baseline Dimension 2x - 7 occurring [either two or] four times in a row, nature seems to dictate, at the very least, perpetual twin **P** or one other non-twin **P** occurrences is inevitable.

We dissect Dimension YYYY unique signal for twin **P** appearances: Initial two CFS Dimensions YY components of YYYY represent "first" **P** component of twin **P** pairing. Last two Dimensions YY components of YYYY signifying appearance of "second" **P** component of twin **P** pairing is also the initial first-two-element component of full CFS Dimensions representation for "first" **P** component of following non-twin **P** pairing. Twin **P** are uniquely represented by repeating *single* type Dimension 2x - 7. In all other 'higher order' **P** pairings (with prime gaps $\geqslant$ 4), they require *multiple* types Dimension representation. There is qualitative aspect association of *single* type Dimension representation for twin **P** resulting in "less colorful" Plus Gap 2 Composite Number *Continuous Law* as opposed to *multiple* types Dimension representation for all other 'higher order' **P** pairings resulting in "more colorful" Plus-Minus Gap 2 Composite Number *Alternating Law*. 'Gap 2 Composite Number' occurrences in both Laws on finite scale are (directly) observed in Figure 8 & Table 2 for x = 2 to 64, and on infinite scale are (indirectly) deduced using logical arguments for all x values.

We endow all "Dimensions" with exponent of '1' for perusal in on-going mathematical arguments. $\mathbf{P}_1 = 2$ is represented by CFS as Dimension $(2x - 4)^1$ (with both prime gap & CFS cardinality = 1); $\mathbf{P}_2 = 3$ is represented by CFS as Dimensions $(2x - 5)^1$ & $(2x - 7)^1$ (with both prime gap & CFS cardinality = 2); $\mathbf{P}_3 = 5$ is represented by CFS Dimension $(2x - 7)^1$ & $(2x - 7)^1$ (with both prime gap & CFS cardinality = 2), etc.

Proposition 9.1. Let Case 1 be Completely Predictable **E** & **O** pairing and Case 2 be Incompletely Predictable **P** & **C** pairing. Furthermore, let Case 1 and Case 2 be independent of each other. Then for any given x value, there exist grand total number of Dimensions [Complexity] such that it exactly equal to either two combined subtotal number of Dimensions [Complexity] to precisely represent **E** & **O** in Case 1, or combined subtotal number of Dimensions [Complexity] to precisely represent **P** & **C** & Number '1' in Case 2.

Proof. N is directly constituted from either combined **E** & **O** in Case 1 or combined **P** & **C** & Number '1' in Case 2 – Number '1' is neither **P** nor **C**. Correctly designated infinitely many CFS of Dimensions used to represent combined **E** & **O** in Case 1 and combined **P** & **C** & Number '1' in Case 2 must also directly and proportionately be representative of relevant **N** arising from combined subtotal of **E** & **O** in Case 1 and from combined subtotal of **P** & **C** & Number '1' in Case 2. *The proof is now complete for Proposition 9.1*□.

Proposition 9.2. Let Case 1 be Completely Predictable **E** & **O** pairing and Case 2 be Incompletely Predictable **P** & **C** pairing. Furthermore, let Case 1 and Case 2 be independent of each other. Part I: For any given x value apart from x = 1 value in Case 1 and x = 1, 2, and 3 values in Case 2; Dimension $(2x - N)^1$ [Complexity] representations of all Completely Predictable **E** & **O** in Case 1 and all Incompletely Predictable **P** & **C** & Number '1' in Case 2 are such that they are given by N = 4 in Case 1 and by N ⩾ 7 in Case 2. Part II: Odd **P** obeys 'Plus-Minus Composite Gap 2 Number Alternating Law' for prime gaps ⩾ 4 and 'Plus Composite Gap 2 Number Continuous Law' for prime gap = 2.

Proof. Apart from first Dimension $(2x - 2)^1$ representation in **E** & **O** pairing in Case 1 and first three Dimension $(2x - 2)^1$, Dimension $(2x - 4)^1$ and Dimension $(2x - 5)^1$ representations in **P** & **C** pairing in Case 2; possible N value in Dimension $(2x - N)^1$ representation are shown to be (constantly) maximal 4 for Case 1 and (variably) minimal 7 for Case 2. For Case 2, we again note Dimension $(2x - 2)^1$ to (validly) represent Number '1' which is neither **P** nor **C**. These nominated Dimensions simply represent possible (constant) baseline "2x - 4" Grand-Total Gaps as per Proposition 8.7 for Case 1 & (variable) baseline "2x - 7" Grand-Total Gaps as per Proposition 8.8 for Case 2. Note that all CFS of Dimensions that can be used to precisely represent combined **E** & **O** in Case 1 will persistently consist of same [solitary] Dimension $(2x - 4)^1$ after first Dimension $(2x - 2)^1$. Perpetual repeated deviation of N values away from N = 7 (minimum) in Case 2 is simply representing infinite magnitude of **P** & **C**. *The proof is now complete for Part I of Proposition 9.2*□.

Derived Dimensions will comply with Incompletely Predictable property as explained using **P** '61'. At Position x = 61 equating to $\mathbf{P}_{18} = 61$, it is represented by CFS Dimensions $(2x - 7)^1$, $(2x - 7)^1$, $(2x - 8)^1$, $(2x - 9)^1$, $(2x - 10)^1$ & $(2x - 11)^1$ (with both prime gap & CFS cardinality = 6). This representation indicates an "unknown but correct" **P** with prime gap = 6 when we intentionally conceal full information '61' = 31st **O** = 18th **P** with prime gap = 6. But to arrive at this representation requires calculations of all preceding CFS Dimensions thus manifesting hallmark Incompletely Predictable property of CFS Dimensions.

Overall sum total of individual CFS Dimensions required to represent every **P** is infinite in magnitude as |all **P**| = $\aleph_0$. Standalone Dimensions YY groupings [representing signals for "higher order" non-twin **P** appearances] &/or as front Dimensions YY (sub)groupings [which by itself is fully representative of twin **P** as Dimensions YYYY appearances] need to recur on an indefinite basis. Then twin **P** and "higher order" cousin **P**, sexy **P**, etc should aesthetically all be infinite in magnitude because (respectively) they regularly and universally arise as part of Dimension YYYY and Dimension YY appearances. An isolated **P** is defined as a **P** such that neither **P** - 2 nor **P** + 2 is **P**. In other words, isolated **P** is not part of a twin **P** pair. Example 23 is an isolated **P** since 21 and 25 are both **C**. Then repeated inevitable presence of Dimension YY grouping is nothing more than indicating repeated occurrences of isolated **P**. This constitutes another view on Dimension YY.

CIS of Gap 1 Composite Numbers are fully associated with non-twin **P** as they eternally occur in between any two consecutive non-twin **P**. CIS of Gap 2 Composite Numbers are (i) fully associated with twin **P** as they are eternally present in between any twin **P** pair, and (ii) partially associated with non-twin **P** as they are eternally present alternatingly or intermittently in between any two consecutive non-twin **P**. Then (i) Gap 1 Composite Numbers do not have valid representation by **E** prime gap = 2, and (ii) Gap 2 Composite Numbers have valid representations by all **E** prime gaps = ["consistently" only for] 2, ["inconsistently" for each of] 4, 6, 8, 10,.... This is an alternative view on **P** from perspective of CFS composite gaps [instead of CIS prime gaps] with intrinsic patterns having *alternating presence* and *absence* of Gap 2 Composite Numbers associated with every CFS Dimensions representations of **P** with prime gaps ⩾ 4, viz. 'Plus-Minus Gap 2 Composite Number Alternating Law'. CFS Dimensions representations of Twin **P** are always associated with Gap 2 Composite Numbers, viz. 'Plus Gap 2 Composite Number Continuous Law'. Examples for both Laws: A twin **P** (prime gap = 2) in its unique CFS Dimensions format always has Gap 2 Composite Numbers in a [constant] pattern. A cousin **P** (prime gap = 4) in its unique CFS Dimensions format always has two Gap 1 Composite Numbers & then one Gap 2 Composite Number [combined] pattern *alternating* with three consecutive Gap 1 Composite Numbers [non-combined] pattern. From this simple observation alone, we deduce we can generate an infinite magnitude of **C** from each composite gaps 1 & 2. Gap 2 Composite Numbers *alternating* pattern behavior in cousin **P** will not hold true unless twin **P** & all other non-cousin **P** are infinite in magnitude and integratedly supplying essential "driving mechanism" to eternally sustain this Gap 2 Composite Numbers *alternating* pattern behavior in cousin **P**. Thus we establish twin **P** and cousin **P** in their CFS Dimensions formats are CIS intertwined together when depicted using **C** with composite gaps = 1 & 2 with each supplying their own peculiar (infinite) share of associated Gap 2 Composite Numbers [thus contributing to overall pool of Gap 2 Composite Numbers].

An inevitable statement in relation to "Gap 2 Composite Numbers pool contribution" based on above reasoning: At the bare minimum, *either* twin **P** *or* at least one of non-twin **P** must be infinite in magnitude. An inevitable impression: All generated subsets of **P** from 'small gaps' [of 2 & 4] and 'large gaps' [of ⩾ 6] alike should each be CIS thus allowing true uniformity in **P** distribution. Again we see in Table 2 above depicting **P-C** data for x = 2 to 64 that, for instance, **P** with prime gap = 6 must also persistently have this 'last-place' Gap 2 Composite Numbers intermittently appearing in certain rhythmic *alternating* patterns, thus complying with Plus-Minus Gap 2 Composite Number Alternating Law. This CFS Dimensions representation for **P** with prime gaps = 6 will again generate their infinite share of associated Gap 2 Composite Numbers to contribute to this pool. The presence of this last-place Gap 2 Composite Numbers in various alternating pattern in appearances & non-appearances must *self-generatingly* be similarly extended in a mathematically consistent fashion *ad infinitum* to all other remaining infinite number of prime gaps [which were not discussed in details above]. *The proof is now complete for Part II of Proposition 9.2*□.

10. Rigorous Proofs now named as Polignac's and Twin prime hypotheses

The proofs on lemmas and propositions from previous section supply all necessary evidences to fully support Theorem Polignac-Twin prime I to IV below thus depicting proofs for Polignac's and Twin prime conjectures in a rigorous manner. Gap 1 Composite Numbers do not have valid representation by **E** prime gap = 2, and Gap 2 Composite Numbers have valid representations by all **E** prime gaps = ["consistently" only for] 2, ["inconsistently" for each of] 4, 6, 8, 10,.... Plus-Minus Gap 2 Composite Number Alternating Law confirms that Gap 2 Composite Numbers present in each **P** with prime gaps ≥ 4 situation must appear as some sort of "rhythmic patterns of alternating presence and absence" for Gap 2 Composite Numbers. Twin **P** with prime gap = 2 obeying Plus Gap 2 Composite Number Continuous Law can be understood as special situation of "(non-)rhythmic patterns with continual presence" for relevant Gap 2 Composite Numbers.

In 1849 when French mathematician Alphonse de Polignac (1826 - 1863) was admitted to Polytechnique, he made what is known as Polignac's conjecture which relates complete set of odd **P** to all **E** prime gaps. Made earlier by de Polignac in 1846, Twin prime conjecture which relates twin prime numbers to prime gap = 2, is nothing more than a subset of Polignac's conjecture.

Theorem Polignac-Twin prime I. Incompletely Predictable prime numbers $\mathbf{P}_n$ = 2, 3, 5, 7, 11, ..., ∞ or composite numbers $\mathbf{C}_n$ = 4, 6, 8, 9, 10, ..., ∞ are CIS with overall actual location [but not actual positions] of all prime or composite numbers accurately represented by complex algorithm involving prime gaps G_{Pi} viz. $\mathbf{P}_{n+1} = 2 + \sum_{i=1}^{n} G_{Pi}$ or involving composite gaps G_{Ci} viz. $\mathbf{C}_{n+1} = 4 + \sum_{i=1}^{n} G_{Ci}$ whereby prime & composite numbers are symbolically represented here with aid of 'n' notation instead of usual 'i' notation; and i & n = 1, 2, 3, 4, 5, ..., ∞. Number '2' in first algorithm represents $\mathbf{P}_1$, the very first (and only even) **P**. Number '4' in second algorithm represent $\mathbf{C}_1$, the very first (and even) **C**.

Proof. We treat above algorithms as unique mathematical objects looking for key intrinsic properties and behaviors. Each **P** or **C** is assigned a unique prime or composite gap. Absolute number of **P** or **C** and (thus) prime or composite gaps are infinite in magnitude. As original formulae containing all **P** or **C** by themselves (viz. without supplying prime or composite gaps as "input information" to generate **P** or **C** as "output complexity"), these algorithms intrinsically incorporate overall actual location [but not actual positions] of all **P** or **C**. *The proof is now complete for Theorem Polignac-Twin prime I*□.

Theorem Polignac-Twin prime II. Set of prime gaps G_{Pi} = 2, 4, 6, 8, 10, ..., ∞ is infinite in magnitude whereby these prime gaps accurately and completely represented by Dimensions $(2x - 7)^1$, $(2x - 8)^1$, $(2x - 9)^1$, ..., $(2x - \infty)^1$ must satisfy Information-Complexity conservation in a consistent manner.

Proof. Part I of Proposition 9.2 proved all **P** are represented by Dimension $(2x - N)^1$ with N ≥ 7 for any given x value (except for x = 2 & 3 values). Note that although x = 1 is neither **P** nor **C**, it is validly represented by Dimension $(2x - 2)^1$. If each **P** is endowed with a specific prime gap value, then each such prime gap must [via logical mathematical deduction] be represented by Dimension $(2x - N)^1$. We advocate this nominated method of prime gap representation using Dimensions be [purportedly] the only way to achieve Information-Complexity conservation. The preceding mathematical statements are correct as there is a unique prime gap value associated with each **P**. Proposition 10.1 below based on principles from Set theory provides further supporting materials that prime gaps are infinite in magnitude. *The proof is now complete for Theorem Polignac-Twin prime II*□.

Theorem Polignac-Twin prime III. To maintain Dimensional analysis (DA) homogeneity, those Dimensions $(2x - N)^1$ from Theorem Polignac-Twin prime II must contain eternal repetitions of well-ordered sets constituted by Dimensions $(2x - 7)^1$, $(2x - 8)^1$, $(2x - 9)^1$, $(2x - 10)^1$, $(2x - 11)^1$, ..., $(2x - \infty)^1$.

Proof. This Theorem is stated in greater details as "To maintain DA homogeneity, those aforementioned [endowed with exponent 1] Dimensions $(2x - N)^1$ from Theorem Polignac-Twin prime II must repeat themselves indefinitely in following specific combinations – (i) Dimension $(2x - 7)^1$ only appearing as twin [two-times-in-a-row] and quadruplet [four-times-in-a-row] sequences, and (ii) Dimensions $(2x - 8)^1$, $(2x - 9)^1$, $(2x - 10)^1$, $(2x - 11)^1$,..., $(2x - \infty)^1$ appearing as progressive groupings of **E** 2, 4, 6, 8, 10,..., ∞." To accommodate the only even **P** '2', exceptions to this DA homogeneity compliance will expectedly occur right at beginning of **P** sequence – (i) one-off appearance of Dimensions $(2x - 2)^1$, $(2x - 4)^1$ and $(2x - 5)^1$ and (ii) one-off appearance of Dimension $(2x - 7)^1$ as a quintuplet [five-times-in-a-row] sequence which is equivalent to (eternal) non-appearance of Dimension $(2x - 6)^1$ at x = 4. [We again note Dimension $(2x - 2)^1$ validly represent Number '1' which is neither **P** nor **C**.] These sequentially arranged sets are CFS whereby from x = 11 onwards, each set always commence initially as 'baseline' Dimension $(2x - 7)^1$ at x = **O** values and always end with its last Dimension at x = **E** values. Each set also have varying cardinality with values derived from all **E**; and correctly combined sets always intrinsically generate two infinite sets of **P** and, by default, **C** in an integrated manner. Our Theorem Polignac-Twin prime III simply represent a mathematical summary derived from Section 8 & 9 of all expressed characteristics of Dimension $(2x - N)^1$ when used to represent **P** with intrinsic display of DA homogeneity. See Proposition 10.2 below for further details on DA aspect. *The proof is now complete for Theorem Polignac-Twin prime III*□.

Theorem Polignac-Twin prime IV. Aspect 1. The "quantitive" aspect to existence of both prime gaps and their associated prime numbers as sets of infinite magnitude will be shown to be correct by utilizing principles from Set theory. Aspect 2. The "qualitative" aspect to existence of both prime gaps and their associated prime numbers as sets of infinite magnitude will be shown to be correct by 'Plus-Minus Gap 2 Composite Number Alternating Law' and 'Plus Gap 2 Composite Number Continuous Law'.

Proof. Required concepts from Set theory involve cardinality of a set with its 'well-ordering principle' application. Supporting materials for these concepts based on 'pigeonhole principle' in relation to Aspect 1 are outlined in Proposition 10.1 below. 'Plus-Minus Gap 2 Composite Number Alternating Law' is applicable to all **E** prime gaps [apart from first **E** prime gap = 2 for twin primes]. The prime gap = 2 situation will obey 'Plus Gap 2 Composite Number Continuous Law'. These Laws are essentially Laws of Continuity inferring underlying intrinsic driving mechanisms that enables infinity magnitude association for both prime gaps & prime numbers to co-exist. By the same token, these Laws have the important implication that they must be applicable to those relevant prime gaps on an perpetual time scale. Supporting materials in relation to Aspect 2 are found in Proposition 9.2 above. *The proof is now complete for Theorem Polignac-Twin prime IV*□.

We note two mutually inclusive conditions: Condition 1. Presence of all Dimensions that repeat themselves on an indefinite basis and with exponent of '1' will give rise to complete sets of **P** & **C** ["DA-wise one & only one mathematical possibility argument" associated with inevitable *de novo* DA homogeneity], and Condition 2. Presence of any Dimension(s) that do not repeat itself (themselves) on an indefinite basis or with exponent other than '1' will give rise to incomplete set of **P** & **C** or incorrect set of non-**P** & non-**C** ["DA-wise mathematical impossibility argument" associated with inevitable *de novo* DA non-homogeneity]. When met, these two conditions will fully support the point that CFS Dimensions representations of **P** & **C** [with respective prime & composite gaps] are totally accurate. Condition 1 reflect proof from Theorem Polignac-Twin prime III above as all **P** & **C** are associated with DA homogeneity when their Dimensions are endowed with exponent of '1'. Condition 2 invoke corollary on inevitable appearance of incomplete **P** or **C** or non-**P** or non-**C** [associated with DA non-homogeneity] being tightly incorporated into this mathematical framework. See Propositions 10.1 and 10.2, and Corollary 10.3 below for supporting materials on DA homogeneity & non-homogeneity.

We analyze **P** (& **C**) in terms of (i) measurements based on cardinality of CIS and (ii) pigeonhole principle which states that if n items are put into m containers, with n>m, then at least one container must contain more than one item. We note that ordinality of all infinite **P** (& **C**) is "fixed" implying that each one of the infinite well-ordered Dimension sets conforming to CFS type as constituted by Dimensions $(2x - 7)^1$, $(2x - 8)^1$, $(2x - 9)^1$, $(2x - 10)^1$, $(2x - 11)^1$, ..., $(2x - \infty)^1$ on respective gaps for **P** (& **C**) must also be "fixed".

Proposition 10.1. "Even number prime gaps are infinite in magnitude with each even number prime gap generating odd prime numbers which are again infinite in magnitude" is supported by principles from Set theory and two Laws based on Gap 2 Composite Number.

Proof. We validly exclude even **P** '2' here. Let (i) cardinality $T = \aleph_0$ for Set **all** odd **P** derived from **E** prime gaps 2, 4, 6,..., ∞, (ii) cardinality $T_2 = \aleph_0$ for Subset **odd P** derived from **E** prime gap 2, cardinality $T_4 = \aleph_0$ for Subset **odd P** derived from **E** prime gap 4, cardinality $T_6 = \aleph_0$ for Subset **odd P** derived from **E** prime gap 6, etc. Paradoxically $T = T_2 + T_4 + T_6 +... + T_\infty$ equation is valid despite $T = T_2 = T_4 = T_6 =... = T_\infty$ [with well-ordering principle "stating that every non-empty set of positive integers contains a least element" fulfilled by each (sub)set]. But if Subset **odd P** derived from one or more **E** prime gap(s) are finite in magnitude, this will breach $\aleph_0$ 'uniformity' resulting in (i) DA non-homogeneity and (ii) inequality $T > T_2 + T_4 + T_6 +... + T_\infty$. In language of pigeonhole principle "stating that if n items are put into m containers with n>m, then at least one container must contain more than one item", residual **odd P** (still CIS in magnitude) not accounted for by CFS-type **E** prime gap(s) will have to be [incorrectly] contained in one (or more) of composite gap(s). These arguments using cardinality constitute proof that (i) **E** prime gaps and (ii) odd **P** generated from each **E** prime gap, must all be CIS. *The proof [on "quantitative" aspect] is now complete for Proposition 10.1*□.

Complete set of **P** is represented by Dimensions $(2x - N)^1$. Table 2 & Figure 8 on **PC** finite scale mathematical landscape depict perpetual repeating features used in "qualitative" statements supporting (i) Plus-Minus Gap 2 Composite Number Alternating Law (stated as **C** with composite gaps = 2 present in each of **P** with prime gaps ≥ 4 situation must be observed to appear as some sort of rhythmic patterns of alternating presence and absence of this type of **C**), and (ii) Plus Gap 2 Composite Number Continuous Law (stated as **C** with composite gaps = 2 continual appearances in each of (twin) **P** with prime gap = 2 situation). Plus-Minus Gap 2 Composite Number Alternating Law has built-in intrinsic mechanism to automatically generate all prime gaps ≥ 4 in a mathematically consistent *ad infinitum* manner. Plus Gap 2 Composite Number Continuous Law has built-in intrinsic mechanism to automatically generate prime gap = 2 appearances in a mathematically consistent *ad infinitum* manner. *The proof [on "qualitative" aspect] is now complete for Proposition 10.1*□.

Proposition 10.2. The presence of Dimensional analysis homogeneity will always result in

Book 2 Three Open Problems by Riemann and Polignac

correct and complete set of prime (and composite) numbers.

Proof. DA homogeneity is completely dependent on all Dimensions being consistently endowed with exponent '1'. As all **P** (& **C**) are "fixed", we deduce from Figure 8 & Table 2 that there is one (& only one) way to represent Information-Complexity conservation using our defined Dimensions. Thus, there is one (& only one) way to depict all **P** (& **C**) using these Dimensions in a self-consistent manner and this is achieved with the one (& only one) DA homogeneity possibility. *The proof is now complete for Proposition 10.2*□.

Corollary 10.3. The presence of Dimensional analysis non-homogeneity will always result in incorrect and/or incomplete set of prime (and composite) numbers.

Proof. For optimal clarity, we endow all Dimensions with exponent '1' depicted as $(2x - 7)^1$, $(2x - 8)^1$, $(2x - 9)^1$, $(2x - 10)^1$, $(2x - 11)^1$,..., $(2x - \infty)^1$. Proposition 5.2 equates DA homogeneity with correct & complete set of **P** (& **C**). There are "more than one" DA possibilities when, for instance, a particular [first] term from $(2x - 7)^0$, $(2x - 8)^1$, $(2x - 9)^1$,..., $(2x - \infty)^1$ "terminates" prematurely and does not perpetually repeat [with loss of continuity]. There are intuitively two 'broad' DA possibilities here; namely, (one) DA homogeneity possibility and (one) DA non-homogeneity possibility – Dimension $(2x - 7)^0$ [= 1] with its exponent arbitrarily set as '0' against-all-trend in this case. Thus Dimension $(2x - 7)^1$ that stop recurring at some point in **P** (or **C**) sequence may cause well-ordered CFS sets from progressive groupings of [**E**] 2, 4, 6, 8, 10,..., ∞ for Dimensions $(2x - 8)^1$, $(2x - 9)^1$, $(2x - 10)^1$, $(2x - 11)^1$,..., $(2x - \infty)^1$ to stop existing (and ultimately for sequential **P** (or **C**) to stop appearing) at that point with ensuing outcome that **P** (or **C**) may overall be incorrectly finite or incomplete in magnitude. Finally also manifesting DA non-homogeneity, a Dimension endowed with fractional exponent values other than '1' such as '$\frac{2}{5}$' or '$\frac{3}{5}$' will result in non-**P** (or non-**C**) [fractional] numbers. *The proof is now complete for Corollary 10.3*□.

Each [fixed] finite scale mathematical landscape "page" as part of [fixed] infinite scale mathematical landscape "pages" for **P** & **C** display Chaos [sensitivity to initial conditions viz. positions of subsequent **P** & **C** are "sensitive" to positions of initial **P** & **C**] and Fractals [manifesting fractal dimensions with self-similarity viz. those aforementioned Dimensions for **P** & **C** are always present, albeit in non-identical manner, for all ranges of $x \geq 2$]. Advocated in another manner, Chaos and Fractals phenomena of those Dimensions for **P** & **C** are always present signifying accurate composition of **P** & **C** in different [predetermined] finite scale mathematical landscape "(snapshot) pages" for **P** & **C** that are self-similar but never identical – and there are an infinite number of these finite scale mathematical landscape "(snapshot) pages". The crucial mathematical step in representing all **P** (& **C**) and prime (& composite) gaps with "Dimensions" based on Information-Complexity conservation allows us to obtain the two Laws based on Gap 2 Composite Numbers and perform DA on these entities. The 'strong' principle argument is DA homogeneity equates to complete set of **P** (& **C**) whereas DA non-homogeneity does not equate to complete set of **P** (& **C**). We could also advocate for a 'weak' principle argument supporting DA homogeneity for **P** (& **C**) in that nature should not "favor" any particular Dimension(s) to terminate and therefore DA non-homogeneity does not, and cannot, exist for **P** (& **C**). Abiding to our advocated convention that 'conjecture' be termed 'hypothesis' once proven; we now label Polignac's & Twin prime conjectures as Polignac's & Twin prime hypotheses.

Creed Odyssey in Mathematics and Medicine series

11. Conclusions

"Mathematics for Incompletely Predictable Problems" is mathematical language and framework describing complex properties present in Incompletely Predictable entities. As seen in Appendix D for Even-Odd number pairing, one could [redundantly] introduce "Mathematics for Completely Predictble Problems" to classify problems involving Completely Predictable entities by defining these problems to be Completely Predictable problems associated with simple properties.

CIS of [Completely Predictable] natural numbers 1, 2, 3, 4, 5, 6, 7,... having CIS of [Completely Predictable] natural gaps 1, 1, 1, 1, 1, 1,... are constituted by three dependent sets of numbers: (i) CIS of [Incompletely Predictable] odd prime numbers 3, 5, 7, 11, 13, 17,... having CIS of [Incompletely Predictable] prime gaps 2, 2, 4, 2, 4,... plus CFS of solitary [Incompletely Predictable] even prime number 2 having CFS of [Incompletely Predictable] prime gap 1 (ii) CIS of [Incompletely Predictable] even and odd composite numbers 4, 6, 8, 9, 10, 12,... having CIS of [Incompletely Predictable] composite gaps 2, 2, 1, 1, 2, 2,.... and (iii) CFS of solitary odd number '1' [neither prime nor composite]. Treated as Incompletely Predictable problems endowed with "meta-properties", we gave relatively elementary proofs on Polignac's & Twin prime conjectures using these relationships by (1) performing Information-Complexity conservation on prime & composite numbers to obtain 'Plus-Minus Gap 2 Composite Number Alternating Law' & 'Plus Gap 2 Composite Number Continuous Law'; and (2) demonstrating DA homogeneity with presence of [solitary] cardinality value $\aleph_0$ occurring in all the [even number prime gap] subsets of prime numbers and in the set of even number prime gaps.

Harnassed properties: (1) Nontrivial zeros and two types of Gram points are [*dependently*] derived from "Axes intercept relationship interface" using Riemann zeta function, or its *proxy* Dirichlet eta function; and (2) Prime and composite numbers are [*dependently*] derived from "Numerical relationship interface" using Sieve of Eratosthenes. Using prime gaps as analogy, there are (for instance) "nontrivial zeros gaps" between any two nontrivial zeros with all these gaps of infinite magnitude being Incompletely Predictable entities. Prime number theorem describes asymptotic distribution of prime numbers among positive integers by formalizing intuitive idea that prime numbers become less common as they become larger through precisely quantifying rate at which this occurs using probability. An indirect spin-off arising out of solving Riemann hypothesis result in absolute and full delineation of prime number theorem. This theorem relates to prime counting function which is usually denoted by $\pi(x)$ with $\pi(x)$ = number of prime numbers $\leqslant$ x. In other words, solving Riemann hypothesis is instrumental in proving efficacy of techniques that estimate $\pi(x)$ efficiently. This confirm "best possible" bound for error ("smallest possible" error) of prime number theorem. In mathematics, logarithmic integral function or integral logarithm li(x) is a special function. Relevant to problems of physics and with number theoretic significance, it occurs in prime number theorem as an estimate of $\pi(x)$ whereby the form of this special function is defined so that li(2) = 0; viz. li(x) $\equiv \int_2^x \frac{du}{\ln u}$ = li(x) - li(2). There are less accurate ways of estimating $\pi(x)$ such as conjectured by Gauss and Legendre at end of 18th century. This is approximately x/$\ln x$ in the sense $\lim_{x \to \infty} \frac{\pi(x)}{x/\ln x} = 1$. Skewes' number is any of several extremely large numbers used by South African mathematician Stanley Skewes as upper bounds for smallest natural number x for which li(x)<$\pi(x)$. These bounds have since been improved by others: there is a crossing near $e^{727.95133}$ but it is not known whether this is the smallest. John Edensor Littlewood, who was Skewes' research supervisor, proved in 1914[8] that there is such a [first] number; and found that sign of diffcrence $\pi(x)$ - li(x) changes infinitely often. This refute all prior numerical evidence that seem to suggest li(x) was always more than $\pi(x)$.

The key point is [100% accurate] $\pi(x)$ mathematical tool being "wrapped around" by [less-than-100% accurate] approximate mathematical tool li(x) infinitely often via this 'sign of difference' changes meant that li(x) is the most efficient approximate mathematical tool. Contrast this with "crude" $x/\ln x$ approximate mathematical tool where values obtained diverge away from $\pi(x)$ at increasingly greater rate when larger range of prime numbers are studied.

In Hybrid method of Integer Sequence classification – see Appendix C, a formula is either non-Hybrid or Hybrid integer sequence. Inequation with two 'necessary' Ratio (R) or equation with one 'unnecessary' R contains non-Hybrid integer sequence. Equation with one 'necessary' R contains Hybrid integer sequence. "In the limit" Hybrid integer sequence approach unique Position X, it becomes non-Hybrid integer sequence for all Positions $\geqslant$ Position X. Kinetic energy (KE) has its endowed units in MJ when m_0 = rest mass in kg and v = velocity in ms^{-1}. In classical mechanics concerning low velocity with v<<c, Newtonian KE = $\frac{1}{2}m_0v^2$. In relativistic mechanics concerning high velocity with v$\geqslant$0.01c, Relativistic KE = $\dfrac{m_0c^2}{\sqrt{1-(v^2/c^2)}} - m_0c^2$. Obtained from the later by binomial approximation or by taking first two terms of Taylor expansion for reciprocal square root, the former approximates the later well at low speed. We arbitrarily divide DA homogeneity into inexact DA homogeneity for ["<100% accurracy"] Newtonian KE and exact DA homogeneity for ["100% accurracy"] Relativistic KE. "In the limit" ['<100% accuracy'] Newtonian KE at low speed approach ['100% accuracy'] Relativistic KE at high speed, we achieve *perfection*. An extremely useful analogy: "In the limit" all three versions of Dirichlet Sigma-Power Laws for Gram[y=0] points, Gram[x=0] points and nontrivial zeros as '*<100% accuracy*' inequations approach *perfection* as '*100% accuracy*' equations, compliance with inexact DA homogeneity becomes compliance with exact DA homogeneity. We note R1 terms in all inequations contain (2n) and (2n-1) 'base quantities' but these are not endowed with fractional exponent (σ+1) as relevant 'unit of measurement'. Treated as Incompletely Predictable problems, we gave relatively elementary proof of Riemann hypothesis and explain the two types of Gram points by using the "meta-properties" of relevant Dirichlet Sigma-Power Laws viz. (1) exact DA homogeneity [occurring when $\sigma = \frac{1}{2}$] in both their equations & inequations and (2) inexact DA homogeneity [occurring when $\sigma \neq \frac{1}{2}$] in both their equations & inequations.

Acknowledgements Huge thanks to Civil Engineer & Mathematician Mr. Rodney Williams, Software Engineer & Mathematician Mr. Tony O'Hagan, and Experts for reviewing this paper.

REFERENCES

1 Hardy, G. H. (1914). Sur les Zeros de la Fonction $\zeta(s)$ de Riemann. *C. R. Acad. Sci. Paris, 158*, 1012-1014. JFM 45.0716.04 Reprinted in (Borwein et al., 2008)

2 Hardy, G. H.; Littlewood, J. E. (1921). The zeros of Riemann's zeta-function on the critical line. *Math. Z., 10* (3-4), 283-317. http://dx.doi:10.1007/BF01211614

3 Abel, N.H. (1823). Solution de quelques problemes a l'aide d'integrales definies. *Magazin Naturvidensk, 1*, 55-68.

4 Plana, G.A.A. (1820). Sur une nouvelle expression analytique des nombres Bernoulliens, propre a exprimer en termes finis la formule generale pour la sommation des suites. *Mem. Accad. Sci. Torino, 25*, 403-418.

5 Furstenberg, H. (1955). On the infinitude of primes. *Amer. Math. Monthly, 62*, (5) 353. http://dx.doi.org/10.2307/2307043

6 Saidak, F. (2006). A New Proof of Euclid's theorem, *Amer. Math. Monthly, 113*, (10) 937. http://dx.doi.org/10.2307/27642094

7 Zhang, Y. (2014). Bounded gaps between primes, *Ann. Math. 179*(3) 1121-1174. http://dx.doi.org/10.4007/annals.2014.179.3.7

8 Littlewood, J. E. (1914). Sur la distribution des nombres premiers. *Comptes Rendus de l'Acad. Sci. Paris, 158*, 1869-1872.

9 Ting, J (2013). *A228186*. The On-Line Encyclopedia of Integer Sequences. https://oeis.org/A228186

10 Noe, T (2004). *A100967*. The On-line Encyclopedia of Integer Sequences. https://oeis.org/A100967

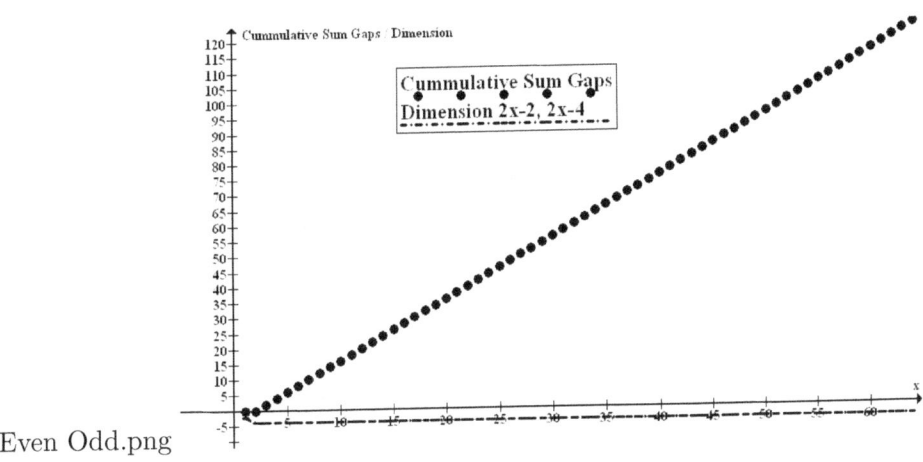

FIGURE 9. Even-Odd mathematical (graphed) landscape using data obtained for x = 1 to 64.

Appendix A. Gram's Law and traditional 'Gram points'

Named after Danish mathematician Jørgen Pedersen Gram (June 27, 1850 – April 29, 1916), traditional 'Gram points' (Gram[y=0] points) are other conjugate pairs values on critical line defined by $Im\{\zeta(\frac{1}{2} \pm \imath t)\} = 0$. Belonging to Incompletely Predictable entities, they obey Gram's Rule and Rosser's Rule with some characteristic properties outlined by our brief exposition below.

Z function is used to study Riemann zeta function on critical line. Defined in terms of Riemann-Siegel theta function & Riemann zeta function by $Z(t) = e^{\imath\theta(t)}\zeta(\frac{1}{2} + \imath t)$ whereby $\theta(t) = \arg(\Gamma(\frac{(2\imath t+1)}{4})) - \frac{\ln \pi}{2}t$; it is also called Riemann-Siegel Z function, Riemann-Siegel zeta function, Hardy function, Hardy Z function, & Hardy zeta function.

The algorithm to compute Z(t) is called Riemann-Siegel formula. Riemann zeta function on critical line, $\zeta(\frac{1}{2} + \imath t)$, will be real when $\sin(\theta(t)) = 0$. Positive real values of t where this occurs are called 'Gram points' and can also be described as points where $\frac{\theta(t)}{\pi}$ is an integer. Real part of this function on critical line tends to be positive, while imaginary part alternates more regularly between positive & negative values. That means sign of Z(t) must be opposite to that of sine function most of the time, so one would expect nontrivial zeros of Z(t) to alternate with zeros of sine term, i.e. when θ takes on integer multiples of π. This turns out to hold most of the time and is known as Gram's Rule (Law) – a law which is violated infinitely often though. Thus Gram's Law is statement [on the manifested property] that nontrivial zeros of Z(t) alternate with 'Gram points'. 'Gram points' which satisfy Gram's Law are called 'good', while those that do not are called 'bad'. A Gram block is an interval such that its very first & last points are good 'Gram

TABLE 3. Even-Odd mathematical (tabulated) landscape using data obtained for x = 1 to 64.
Legend: **E**=even, **O**=odd, Dim = Dimension, Y=Dimension 2x-4.

x	E_i or O_i, Gaps	ΣEO_x-Gaps	Dim	x	E_i or O_i, Gaps	ΣEO_x-Gaps	Dim
1	O1, 2	0	2x-2	33	O17, 2	62	Y
2	E1, 2	0	Y	34	O17, 2	64	Y
3	O2, 2	2	Y	35	O17, 2	66	Y
4	E2, 2	4	Y	36	O17, 2	68	Y
5	O3, 2	6	Y	37	O17, 2	70	Y
6	E3, 2	8	Y	38	O17, 2	72	Y
7	O4, 2	10	Y	39	O17, 2	74	Y
8	E4, 2	12	Y	40	O17, 2	76	Y
9	O5, 2	14	Y	41	O17, 2	78	Y
10	E5, 2	16	Y	42	O17, 2	80	Y
11	O6, 2	18	Y	43	O17, 2	82	Y
12	E6, 2	20	Y	44	O17, 2	84	Y
13	O7, 2	22	Y	45	O17, 2	86	Y
14	E7, 2	24	Y	46	O17, 2	88	Y
15	O8, 2	26	Y	47	O17, 2	90	Y
16	E8, 2	28	Y	48	O17, 2	92	Y
17	O9, 2	30	Y	49	O17, 2	94	Y
18	E9, 2	32	Y	50	O17, 2	96	Y
19	O10, 2	34	Y	51	O17, 2	98	Y
20	E10, 2	36	Y	52	O17, 2	100	Y
21	O11, 2	38	Y	53	O17, 2	102	Y
22	E11, 2	40	Y	54	O17, 2	104	Y
23	O12, 2	42	Y	55	O17, 2	106	Y
24	E12, 2	44	Y	56	O17, 2	108	Y
25	O13, 2	46	Y	57	O17, 2	110	Y
26	E13, 2	48	Y	58	O17, 2	112	Y
27	O14, 2	50	Y	59	O17, 2	114	Y
28	E14, 2	52	Y	60	O17, 2	116	Y
29	O15, 2	54	Y	61	O17, 2	118	Y
30	E15, 2	56	Y	62	O17, 2	120	Y
31	O16, 2	58	Y	63	O17, 2	122	Y
32	E16, 2	60	Y	64	O17, 2	124	Y

points' and all 'Gram points' inside this interval are bad. Counting nontrivial zeros then reduces to counting all 'Gram points' where Gram's Law is satisfied and adding the count of nontrivial zeros inside each Gram block. With this process we do not have to locate nontrivial zeros, and we just have to accurately compute Z(t) to show that it changes sign.

Appendix B. Ratio Study and Inequations

A mathematical equation, containing one or more variables, is a statement that values of two ['left-hand side' (LHS) and 'right-hand side' (RHS)] mathematical expressions is related as equality: LHS = RHS; or as inequalities: LHS < RHS, LHS > RHS, LHS ⩽ RHS, or LHS ⩾ RHS. A ratio is one mathematical expression divided by another. The term 'unnecessary' Ratio (R) for any given equation is explained by two examples: (1) LHS = RHS and with rearrangement, 'unnecessary' R is given by $\frac{LHS}{RHS} = 1$ or $\frac{RHS}{LHS} = 1$; and (2) LHS > RHS and with rearrangement, 'unnecessary' R is given by $\frac{LHS}{RHS} > 1$ or $\frac{RHS}{LHS} < 1$. Consider exponent y ∈ all **R** values & base x ∈ **R**⩾0 values for mathematical expression x^y. Equations such as x^1 = x, x^0 = 1 & 0^y = 0 are all valid. Simultaneously letting both x & y = 0 is an incorrect mathematical action because x^y as function of two-variables is not continuous & is thus undefined at Origin. But if we elect to carry out this "balanced" action [equally] on x & y, we obtain (simple) inequation $0^0 \neq 1$ with associated perpetual obeyance of '=' equality symbol in x^y for all applicable **R** values except when both x & y = 0. The Number '1' value in this inequation is justified by two arguments: I. Limit of x^y value as both x & y tend to zero (from right) is 1 [thus fully satisfying criterion "x^y is right continuous at the Origin"]; and II. Expression x^y is product of x with itself y times [and thus x^0, the "empty product", should be 1 (no matter what value is given to x)].

Mathematical operator 'summation' must obey the law: We can break up a summation across a sum or difference but not across a product or quotient viz, factoring a sum of quotients into a corresponding quotient of sums is an incorrect mathematical action. But if we elect to carry out this action equally on LHS & RHS products or quotients in a suitable equation, we obtain two (unique) 'necessary' R denoted by R1 for LHS and R2 for RHS whereby R1 ≠ R2 relationship will always hold. We define 'Ratio Study' as performing this incorrect [but "balanced"] mathematical action on suitable equation [equivalent to one (non-unique) 'unnecessary' R] to obtain its inequation [equivalent to two (unique) 'necessary' R]. Set $\mathbb{C}$ is a field (but not an ordered field). Thus it is not possible to define a relation between two given (z_1 & z_2) $\mathbb{C}$ as $z_1 < z_2$ since inequality operation here is not compatible with addition and multiplication. Performing Ratio Study to obtain inequations involving $\mathbb{C}$ does not involve defining a relation between two $\mathbb{C}$.

Appendix C. Hybrid method of Integer Sequence classification

Hybrid method of Integer Sequence classification enables meaningful division of all integer sequences into either Hybrid or non-Hybrid integer sequences. Our exotic A228186 integer sequence[9] was published on The On-line Encyclopedia of Integer Sequences website in 2013. It is the first ever [infinite length] Hybrid integer sequence synthesized from Combinatorics Ratio. In 'Position i' notation, let i = 0, 1, 2, 3, 4, 5,..., ∞ be complete set of natural numbers. A228186 "Greatest k > n such that ratio R < 2 is a maximum rational number with R = $\frac{CombinationsWithRepetition}{CombinationsWithoutRepetition}$" is equal to [infinite length] non-Hybrid (usual garden-variety) integer sequence A100967[10] except for finite 21 'exceptional' terms at Positions 0, 11, 13, 19, 21, 28, 30, 37, 39, 45, 50, 51, 52, 55, 57, 62, 66, 70, 73, 77, and 81 with their values given by relevant A100967 terms plus 1. The first 49 terms [from Position 0 to Position 48] of A100967 "Least k such that binomial(2k+1, k-n) ⩾ binomial(2k, k)" are listed below: 3, 9, 18, 29, 44, 61, 81, 104, 130, 159, 191, 225, 263, 303, 347, 393, 442, 494, 549, 606, 667, 730, 797, 866, 938, 1013, 1091, 1172, 1255, 1342, 1431, 1524, 1619, 1717, 1818, 1922, 2029, 2138, 2251, 2366, 2485, 2606, 2730, 2857, 2987, 3119, 3255, 3394, and 3535. For those 21 'exceptional' terms: at Position 0, A228186 (= 4) is given by A100967 (= 3) + 1; at Position 11, A228186 (= 226) is given by A100967 (= 225) + 1; at Position 13, A228186 (= 304) is given by A100967 (= 303) + 1; at Position 19, A228186 (= 607) is given by A100967 (= 606) + 1; etc. Useful concept: Commencing from Position 0 onwards "in the limit" that this Position approaches 82, A228186 Hybrid integer sequence becomes (& is identical to) A100967 non-Hybrid integer sequence for all Positions⩾82.

Appendix D. Tabulated and graphical data on Even-Odd mathematical landscape

We tabulate (in Table 3) and graph (in Figure 9) [Completely Predictable] **E-O** mathematical landscape for x = 1 to 64. Involved Dimensions are 2x - 2 & 2x - 4 with Y denoting Dimension 2x - 4 for visual clarity. This mathematical landscape of Dimension 2x - 4 (except for first and only Dimension 2x - 2) will intrinsically incorporate **E** & **O** in an integrated manner. Except for first **O**, all Completely Predictable **E** & **O** and all their associated gaps are represented by countable finite set of [single] Dimension 2x - 4. Dimensions 2x - 2 & 2x - 4 are symbolically represented by -2 & -4 with 2x - 4 displayed as 'baseline' Dimension whereby Dimension trend (Cumulative Sum Gaps) must reset itself onto this (Grand-Total Gaps) 'baseline' Dimension after initial Dimension 2x - 2 on a permanent basis. Graphical appearances of Dimensions symbolically represented by two negative integers are Completely Predictable with both Even-$\pi(x)$ and Odd-$\pi(x)$ becoming larger at a constant rate. There is a complete absence of Chaos and Fractals phenomena.

Definitive derivation of data in Table 3 is illustrated by two examples for position x = 31 & 32. For i & x ∈ 1, 2, 3, ..., ∞; ΣEO_x-Gap = ΣEO_{x-1}-Gap + Gap value at E_{i-1} or Gap value at O_{i-1} whereby (i) E_i or O_i at position x is determined by whether relevant x value belongs to **E** or **O**, and (ii) both ΣEO_1-Gap and ΣEO_2-Gap = 0. Example, for position x = 31: 31 is **O** (O16). Our desired Gap value at O15 = 2. Thus ΣEO_{31}-Gap (58) = ΣEO_{30}-Gap (56) + Gap value at O15 (2). Example, for position x = 32: 32 is **E** (E16). Our desired Gap value at E15 = 2. Thus ΣEO_{32}-Gap (60) = ΣEO_{31}-Gap (58) + Gap value at E15 (2).

Creed Odyssey in Mathematics and Medicine series

Chapter 1 The Big Picture of Fundamental Laws

"There are inherent Black and White Laws with complete accuracy applicable to Nonliving Things but only Black and White–like Laws with incomplete accuracy applicable to Living Things." By Dr. John Ting (June 1, 2019).

Above is my quote using Black and White (B&W) concept when applied to "Elementary" Nonliving Things and "Emergent" Living Things. Mathematical-based proofs for Nonliving Things with simple or complex Elementary problems must be B&W correct (with absolute 100% certainty). Probability-based proofs for Living Things with simple or complex Emergent problems can only be B&W–like correct (with arbitrarily chosen level of statistical significance of less than and thus never equal to 100% certainty).

My two main research papers in 2019 on Riemann hypothesis, Polignac's and Twin prime conjectures were submitted to a mathematical journal. These can be accessed as "original version" viXra papers listed as (Ting J., Solving Incompletely Predictable problem Riemann hypothesis with Dirichlet Sigma-Power Law, April 2019) and (Ting J., Solving Incompletely Predictable problems Polignac's and Twin Prime conjectures using Information-Complexity conservation, April 26, 2019). These two papers are now combined together as one overall paper "Mathematics for Incompletely Predictable Problems: Riemann zeta function and Sieve of Eratosthenes" (November 5, 2019) which is reproduced at the beginning of this book (and every other books in this series).

Errata Notice: There are mathematical errors present in the first and second paper as relevant equations which were [incorrectly] treated as 'equations' instead of being [correctly] treated as 'inequations'. These papers are corrected, refined and amalgamated together as (Ting J., Solving Incompletely Predictable problem Riemann hypothesis with Dirichlet Sigma-Power Law http://vixra.org/pdf/1903.0483v6.pdf, April 12, 2019) with all equations, inequations and mathematical arguments peer reviewed to be correct and complete.

We compare for similarities and contrast for differences between Riemann hypothesis (RH) and Polignac's and Twin prime conjectures (P&TPC) below.

Firstly, RH and P&TPC are *sine qua nons* classified as Incompletely Predictable problems with their rigorous proofs only possible when acknowledged and treated as such.

Secondly, to solve Completely Predictable problems require the simplicity of deriving their rigorous proofs based on simple properties whereas to solve Incompletely Predictable problems require the complexity of deriving their rigorous proofs based on complex properties ("meta-properties"). For Incompletely Predictable problems, the [few] complex properties are derived from the [many] underlying simple properties. As opposed to Completely Predictable problems such as dealing with even number gaps (= 2) or odd number gaps (= 2) endowed with simple properties which are easy to solve, dealing with RH and P&TPC which are Incompletely Predictable problems endowed with complex properties are mind-boggling hard to solve. RH and P&TPC are respectively connected with Incompletely Predictable entities nontrivial zeros (directly) derived from Riemann zeta function and prime numbers (directly) derived from Sieve of Eratosthenes. The act of explaining closely related Incompletely Predictable entities of the two types of Gram points which are (directly) derived [dependently] from Riemann zeta function should logically be classified as Incompletely Predictable problems. The Incompletely Predictable entities of composite numbers which are (indirectly) derived [dependently] from Sieve of Eratosthenes should logically be incorporated together with prime numbers to help solve P&TPC.

Thirdly, the two sets of prime and composite numbers exist at 'Numerical relationship interface' with (solitary) "outlier" even prime number '2'; and the three sets of nontrivial zeros (or Gram[x=0,y=0] points), Gram[y=0] points (or 'traditional' Gram points) and Gram[x=0] points exist at 'Axes intercept relationship interface' with (solitary) "outlier" negative Gram[y=0] point.

Fourthly, deep seated connections exist between Riemann zeta function, $\zeta(s)$, and complete Set all prime numbers 2, 3, 5, 7, 11, 13,... [but not complete Subsets of prime numbers with each uniquely derived from prime gaps 1, 2, 4, 6, 8, 10,...]. The equivalent Euler product formula from Equation 1 in (Ting J., April 12, 2019) with product over all prime numbers [instead of summation over natural numbers] can also be used to represent $\zeta(s)$. Thus Set all prime numbers is intrinsically "inscribed" in $\zeta(s)$. Prime number theorem, fully delineated by prime counting function [denoted by $\pi(x)$], describes asymptotic distribution of all prime numbers among positive integers by formalizing intuitive idea that prime numbers become less common as they become larger through precisely quantifying rate at which this occurs using probability. Note: we must instead use Dirichlet eta function, $\eta(s)$, [which is the proxy function for $\zeta(s)$] to solve RH. Solving RH is instrumental in proving efficacy of techniques that estimate $\pi(x)$ efficiently thus confirming "best possible" bound for error ("smallest possible" error) of this theorem.

Fifthly, solving RH involves rigorously proving the complete Set nontrivial zeros [of known infinite magnitude] to be located on so-called critical line whereas solving P&TPC involves rigorously proving the existence of complete Set all prime numbers [of known infinite magnitude] to be constituted by Subsets of prime numbers [each proposed to be of infinite magnitude] uniquely derived from all even number prime gaps [proposed to be of infinite magnitude]. RH deals with 'Set' whereas P&TPC deals with 'Subsets'. Note that Polignac's conjecture concerns Subsets of prime numbers derived from all even number prime gaps 2, 4, 6, 8, 10,... but Twin prime conjecture concerns Subset of prime numbers derived only from even number prime gap 2. Thus the later conjecture is intrinsically just part of the former conjecture. From Set theory, the Sets and Subsets of prime numbers will always comply with 'well-ordering Principle' (which states that every non-empty set of positive integers contains a least element) and 'pigeonhole principle' (which states that if n items are put into m containers with n¿m, then at least one container must contain more than one item).

We now provide a Hierarchical Classification for Elementary-Emergent Fundamental Laws (EEFL). Implied by the definition for 'Fundamental Laws', then EEFL must by default be perfectly applicable to Terrestrial human beings on planet Earth (endowed with advanced civilization) and also Extraterrestrial alien beings on some hypothetical remote planet (endowed with super-advanced civilization). Thus one could also appropriately coin our Fundamental Laws as the Extraterrestrial-Terrestrial EEFL.

In order of increasing complexity, we have the following Laws:

Creed Odyssey in Mathematics and Medicine series

Law I: Simple Elementary Fundamental Law for "simple" Nonliving Things with simple properties
Law II: Complex Elementary Fundamental Law for "complex" Nonliving Things with complex properties
Law III: Simple Emergent Fundamental Law for "simple" Living Things with simple properties
Law IV: Complex Emergent Fundamental Law for "complex" Living Things with complex properties

Solving Completely Predictable and Inompletely Predictable problems:

Solving Completely Predictable problems in both Simple 'Nonliving' Elementary and 'Living' Emergent cases: Many Simple properties $\longrightarrow$ [Simple Elementary and Emergent Solutions]
Solving Incompletely Predictable problems in both Complex 'Nonliving' Elementary and 'Living' Emergent cases: Many Simple properties $\longrightarrow$ Few Complex properties $\longrightarrow$ [Complex Elementary and Emergent Solutions]

The three types of entities:

Type I Entities	Completely Unpredictable entities
Type II Entities	Completely Predictable entities
Type III Entities	Incompletely Predictable entities

Type I Entities occur purely as totally random physical processes in nature e.g. radioactive decay is a stochastic (random) process occurring at level of single atoms. According to Quantum theory, it is impossible to predict when a particular atom will decay regardless of how long the atom has existed. For a collection of atoms, expected decay rate is characterized in terms of their measured decay constants or half-lives.

The location-based definitions for Type II Entities and Type III Entities are:

Completely Predictable (Type II) Entities: Locationally defined as entities whose position is independently determined by simple calculations using simple equation or algorithm without needing to know related positions of all preceding entities in neighborhood.
Incompletely Predictable (Type III) Entities: Locationally defined as entities whose position is dependently determined by complex calculations using complex equation or algorithm with needing to know related positions of all preceding entities in neighborhood.

Postulated association between Entities and Laws:

Law I is obeyed by Type II Entities e.g. (simple Nonliving Thing) even and odd numbers with even number and odd number gaps.
Law II is obeyed by Type III Entities e.g. (complex Nonliving Thing) nontrivial zeros related to RH, and prime numbers related to P&TPC.
Law III is obeyed by Type II + Type III Entities e.g. (simple Living Thing) human heart as an organ manifesting hemodynamic and electrical properties.
Law IV is obeyed by Type I + Type II + Type III Entities e.g. (complex Living Thing) human brain which is often dubbed "the most complex structure in the universe" manifesting a whole range of neuro-psychological and neuro-psychiatric properties.

Human heart can simplistically be thought of having a "plumbing system" consisting of heart muscle pump, coronary arteries and cardiac valves; and an "electrical system" consisting of specialized heart muscle cells giving rise to pacemakers and electrical conduction pathways & networks.

Human brain is the most complex organ in human body. It produces our every thought, action, memory, feeling and experience of the world. It consists of jelly-like mass of tissue weighing around 1.4 kilograms, and contains a staggering one hundred billion nerve cells (neurons).

The complexity of the connectivity between these cells is mind-blowing with each neuron making contact with thousands or even tens of thousands of others, via tiny structures called synapses. Our brains form about a million new connections per second. Our conscious mind commands and our subconscious mind obeys. Thus, our subconscious mind is an unquestioning servant that works day and night to make our behavior fits a pattern consistent with our emotionalized thoughts, hopes, and desires. The pattern and strength of the connections is constantly changing and no two brains are alike. It is in these changing connections that memories are stored, subconscious mind operate, habits learned and personalities shaped by reinforcing certain patterns of brain activity, and losing others.

Structurally, the human brain contains "grey matter" and "white matter". The grey matter is the cell bodies of the neurons, while the white matter is the branching network of thread-like tendrils called dendrites and axons that spread out from the cell bodies to connect to other neurons. However, the human brain also has another even more numerous type of cell called glial cells. These outnumber neurons about ten times over. Once thought to be support cells, they are now known to amplify neural signals and to be as important as neurons in mental calculations.

In summary, the human brain manifest Natural Intelligence, consciousness, self-awareness, memory; mental illness such as anxiety, depression, schizophrenia; "dark triad" of personality consisting of three negative traits [viz. the tendency to manipulate others (Machiavellianism), seek admiration and special treatment (narcissism), and to be callous and insensitive (psychopathy)]; and "light triad" of personality consisting of three positive traits [viz. the opposite of Machiavellianism (Kantianism), valuing dignity and worth of each individual person (humanism), and believing that people are fundamentally good (Faith in humanity)].

Artificial Intelligence (AI) in Nonliving Things can be regarded as human endeavor to simulate Natural Intelligence in Living Things using powerful computers such as super-computers or quantum computers. DNA is a double helix, while RNA is a single helix. Both have sets of nucleotides that contain genetic information. DNA is a molecule that contains instructions for Living Things to be born, mature, reproduce, and died.

One would commonly concur that there are 'Simple' Living Things such as bacteria without brain and 'Complex' Living Things such as intelligent human with highly developed brain. The dividing line between Living Things and Nonliving Things is that the former is "powered" by DNA with an important implication that Natural Intelligence, consciousness and self-awareness can only be "powered" by DNA [which are organic]. Then by reasonable assumption, properties such as consciousness and self-awareness can never be present in AI created using computers [which are inorganic].

Chapter 2 Religion and Mitochondrion

Creationism versus Evolution debate for Nonliving Things (obeying Law I and Law II) giving rise to Living Things (obeying Law III and Law IV) is compared and contrasted below:

Process of Creationism: Associated with major religions e.g. Islam and Christianity. From the Bible, Adam and Eve was estimated to be created just over 6,000 years ago by world's leading young-earth creationist organizations.
Process of Evolution: Atheists usually believe the Big Bang (when our Universe was created) occur about 13.8 billion years ago. The first true man appeared 13.7998 billion years after the beginning (or about 200,000 years ago).

People from most western countries generally embrace Christianity which is symbolized by the Bible. This is reportedly the biggest bestseller of all time with historically the Bible Creation starting at Before Christ (B.C.) 2000 whereby the earliest Scriptures are handed down from generation to generation orally. Circa B.C. 2000 - 1500 is when the book of Job, perhaps the oldest book of the Bible, was written. The New King James Version (NKJV) of the Bible, as an example, was published in the modern era in [A.D.] 1982. The term anno Domini (A.D.) is Medieval Latin meaning "in the year of the Lord" used to label or number years in the Julian and Gregorian calendars.

To avoid conflicts, humanity must respect the freedom to practice [or not practice] all different religions. How can we reconcile the huge time discrepancy noted above between the process of creationism and evolution? A controversial thought is perhaps the process of "simplified" evolution with natural selection (survival of the fittest) and adaptation as plausible mechanisms occurs in both Simple and Complex Living Things on short, medium and long term scale in past, present and future. Complex Living Things with brains can only arise through creationism. In particular, the complex neuronal brain tissue can only be "created" by God and cannot "evolve" from simple living tissue over the four eons of geologic time scale.

The human genome is the complete set of nucleic acid sequences for humans, encoded as DNA within the 23 chromosome pairs in cell nuclei and in a small DNA molecule found within individual mitochondrion. Mitochondrion is an organelle found in large numbers in most human and non-human living cells in which the biochemical processes of respiration and energy production occur. It has a double membrane, the inner part being folded inwards to form layers (cristae). Population geneticists believe ancestral human population lived somewhere in Africa and started to split up some time after 144,000 years ago [give or take 10,000 years] – the inferred time at which both the mitochondrial and Y chromosome trees make their first branches.

Mitochondria [plural noun] can thus be used to study the detailed Human Family Tree whereby they live inside human cells but outside the nucleus, thus escaping the shuffling of genes that occurs between generations and are passed unchanged from mother to children. In other words, the tiny rings of genetic material in mitochondria are bequeathed only by the egg cell and thus through the maternal line. In principle, all people should have the same string of DNA letters in their mitochondria. In practice, mitochondrial DNA has steadily accumulated changes over the centuries because of copying errors and radiation damage. This has resulted in different set of lineages for mitochondria being descended in different racial groups from particular regions and continents. Example, Europeans belong to a different set of lineages designated H through K and T through X. The split between the two main branches in the European tree suggests that modern humans reached Europe 39,000 to 51,000 years ago.

The following are subjective comments. For Nonliving Things, we would intuitively associate performing calculation $2 + 3 = 3 + 2 = 5$ as a simple case of elementary Completely Predictable problem; and solving RH and P&TPC as complex cases of elementary Incompletely Predictable problems. For Living Things, we intuitively associate analyzing human heart as a simple case of emergent Completely Predictable problem; and analyzing human brain as a complex case of emergent Incompletely Predictable problem. Because mathematical language for describing complex Incompletely Predictable problems in Nonliving Things (such as weather forecasting) and Living Things (such as determining neurophysiology of human memory) are convoluted, we can only ever obtain approximate models of these problems.

FIGURE 10. Image of rings of planet Saturn taken by Cassini spacecraft on June 26, 2016 (used in accordance with NASA Media Usage Guidelines)

All humans are born with an innate desire to explore and ponder the "Unknown". Though Albert Einstein (Born: 14 March 1879, Ulm, Germany. Died: 18 April 1955, Princeton Medical Center, New Jersey, United States) was the most famous scientist of his time, he knew we could never fully understand workings of the world within limitations of human mind. Experiencing the universe as a harmonious whole, he encouraged the use of intuition to solve problems, marvelled at the mystery of God in nature, and applauded the ideals of great spiritual teachers such as Buddha and Jesus.

Albert Einstein's religious views have been widely studied and often misunderstood. Einstein stated that he believed in the pantheistic God of Baruch Spinoza. He did not believe in a personal God who concerns himself with fates and actions of human beings, a view which he described as naïve. He clarified however that, "I am not an atheist", preferring to call himself an agnostic, or a "religious nonbeliever." Einstein also stated he did not believe in life after death, adding "one life is enough for me." He was closely involved in his lifetime with several humanist groups.

In modern times, collaborative human projects in space exploration (such as the stupendously long space flight conducted by NASA's Cassini spacecraft) have provided mankind with glimpses of the splendour and vastness of our Universe. I am fond of Einstein's idea that science is a quest for the secrets of the Old One – his metaphor for the creator of the Universe.

Creed Odyssey in Mathematics and Medicine series

Chapter 3 Beautiful Mathematics versus Sexy Mathematics

Riemann hypothesis was proposed by famous German mathematician Bernhard Riemann (17 September 17, 1826 – July 20, 1866) in 1859. Twin prime and Polignac's conjectures were proposed by French mathematician Alphonse de Polignac (1826 – 1863) in, respectively, 1846 and 1849. In this chapter, I offer my personal opinion with some bold statements on why intractable open [Incompletely Predictable] problems in Number theory of Riemann hypothesis (RH), Polignac's and Twin prime conjectures (P&TPC) have previously not been solved for over 150 years. Chapter 11 above outlining complicated similarities and differences between these open problems will already provide good insight why delay in solving them occur.

Perhaps most mathematicians have a weird sense of humor. My task to explain this delay is simplified using quirky terms 'Beautiful Mathematics' (BM) and 'Sexy Mathematics' (SM) to provide succinct mental pictures thus promoting optimal understanding by the general public.

Beautiful Mathematics (BM): Mental picture for Completely Predictable problems which are easy to solve requiring BM which involves analyzing their (intrinsic) simple properties.
Sexy Mathematics (SM): Mental picture for Incompletely Predictable problems which are difficult to solve requiring SM which involves analyzing their (intrinsic) complex properties.

With the [few] complex properties derived from the underlying [many] simple properties in Incompletely Predictable problems, the caveat is that only through correctly analyzing these complex properties (or "meta-properties") will we ever obtain their complete solutions. Underlying simple properties in Incompletely Predictable problems can be falsely perceived to be complex properties. Complex properties in Incompletely Predictable problems can be hidden away in some subtle manner. Thus actual complex properties are notoriously difficult to correctly decipher with mathematicians frequently barking up the wrong tree in Incompletely Predictable problems. Another mistake is mathematicians utilizing "manifestations" of relevant complex equations or complex algorithms from current Incompletely Predictable problems to compare with near-identical "manifestations" derived from some other seemingly-related Incompletely Predictable problems that were successfully solved in the past; and subsequently [incorrectly] claiming successful proof – this is dubbed by me as 'pseudo-proof'.

In other words for Incompletely Predictable problems, barking up the wrong tree is the equivalent of mathematician [incorrectly] analyzing the "beautiful tree" with simple properties using BM while barking up the right tree is the equivalent of mathematician [correctly] analyzing the "sexy tree" with complex properties using SM. Figuratively speaking, there are many "beautiful trees" choices but only a few "sexy trees" choices. So what actually is this so-called SM for RH and P&TPC? The answer is illustrated using the "mathematical impasse" phenomenon for Completely Predictable problem involving even numbers and their gaps (in Diagram 1) and for Incompletely Predictable problem involving prime numbers and their gaps (in Diagram 2).

Diagram 1. Legend: Even numbers = E, even number gaps = eGap.

E	2	4	6	8	10	12	
eGap		2	2	2	2	2	2

Question: Prove the proposal that even number gaps are always constant and non-varying. Answer: Finite calculations shown in Diagram 1 depict and support even number gaps [= 2] is constant and non-varying but even numbers are infinite in magnitude requiring an infinite number of calculations ("mathematical impasse") in order to show these gaps will always be constant and non-varying.

Book 2 Three Open Problems by Riemann and Polignac

Obtaining rigorous proof then consist of recognizing this as Completely Predictable problem deriving a Completely Predictable 'non-varying' equation for calculating all even numbers which will [intrinsically] contain simple property "all even number gaps = 2". This equation is literally the 'Simple Container' containing all even numbers.

Diagram 2. Legend: Prime numbers = P, prime number gaps = pGap.

P	2		3		5		7		11		13	
pGap		1		2		2		4		2		4

Question: Prove the proposal that apart from first prime number gap [=1] followed by next two consecutive prime number gaps [= 2], prime number gaps are always even numbers and varying. Answer: Finite calculations shown in Diagram 2 depict and support prime number gaps [after the third one] are even numbers and varying but prime numbers are infinite in magnitude requiring an infinite number of calculations ("mathematical impasse") in order to show these gaps will always be even numbers and varying.

Obtaining rigorous proof then consist of recognizing this as Incompletely Predictable problem deriving an Incompletely Predictable 'varying' equation for calculating all prime numbers which will [intrinsically] contain complex property "all prime gaps are even numbers and perpetually varying". This equation is literally the 'Complex Container' containing all prime numbers.

Below is another approach to grasp the concepts of "Incompletely Predictable" versus "Completely Predictable" and "Beautiful Mathematics" versus "Sexy Mathematics".

Why read my book? How Riemann hypothesis (RH) is successfully solved is one of the main reason to justify reading this book. Many current mathematical theorems contain proofs that depend on deriving the correct solution for RH. It is often quoted along the line that "successfully solving RH will immediately result in proving 500 theorems or more at once". One big reason why RH, Polignac's and Twin prime conjectures are notoriously difficult to solve despite their rigorous proofs being literally "right under our nose" is simply because of mathematicians failing to crucially recognize and accept that entities or numbers such as nontrivial zeros and prime numbers associated with these open problems are Incompletely Predictable entities! However, after having recognize and accept these entities as "Incompletely Predictable" [as described next]; it still takes the author a mind-boggling effort to rigorously prove these problems with "correctness" and "completeness" in the sense that the total steps of mathematical arguments must be numerically "100% complete" and mathematically "100% correct" in all relevant submitted proofs under peer review.

What is the meaning of "Incompletely Predictable" versus "Completely Predictable"? The set of Natural numbers 1, 2, 3, 4, 5, 6,.... are made up of the two subsets of Even numbers 2, 4, 6, 8, 10,... and Odd numbers 1, 3, 5, 7, 9, 11,... The two subsets of Even and odd numbers are "Independent" and "Completely Predictable". Example: the next even number after 2,984 (which is the 2,984 / 2 = 1,492nd Even number) is [easily] seen to [independently] be 2,984 + 2 = 2,986 (which is the 2,986 / 2 = 1,493rd Even number). The set of Natural numbers 1, 2, 3, 4, 5, 6,.... are made up of the Number '1' (which is neither Prime nor Composite) plus the two subsets of Prime numbers 2, 3, 5, 7, 11,... and Composite numbers 4, 6, 8, 9, 10, 12,... The two subsets of Prime and Composite numbers are "Dependent" and "Incompletely Predictable". Example: the next sixth prime number 13 [after the fifth prime number 11] has to be [not easily and dependently] computed from "scratch" as: 2 is first prime number, 3 is second prime number, 4 is the first composite number, 5 is the third prime number, 6 is the second composite number, 7 is the fourth prime number, 8 is the third composite number, 9 is the fourth composite number, 10 is the fifth composite number, 11 is the fifth prime number, 12 is the sixth composite number, and finally our desired 13 is the sixth prime number.

How were the proofs for Riemann hypothesis, Polignac's and Twin prime conjectures derived? Using colloquially-speaking "Beautiful Mathematics" (for Completely Predictable problems) versus "Sexy Mathematics" (for Incompletely Predictable problems). Recommend the general public to read the rest of this book to understand more.

"Thanks John! ...Very impressive!" adapted from feedback given by my learned friend Les from Australia in his 2:16 PM June 30, 2019 email reply to my layman explanations on Riemann hypothesis, Polignac's and Twin prime conjectures [as literally given above]. This feedback use, incorporating [first] name identity of person providing it, is approved by Les.

As outlined in (Ting J., April 12, 2019) containing rigorous proof for RH and explaining Gram points, our 'overall' complex properties consist of three variants of Dirichlet Sigma-Power Laws precisely manifesting the required exact and inexact Dimensional analysis homogeneity. These novel Laws are derived from Dirichlet eta function, the proxy function for Riemann zeta function. As outlined in (Ting J., April 26, 2019) containing rigorous proofs for P&TPC, our 'overall' complex properties consist of Plus-Minus Gap 2 Composite Number Alternating Law being precisely obeyed by all even number prime gaps apart from first even number prime gap precisely obeying Plus Gap 2 Composite Number Continuous Law. These Laws are derived using novel research method Information-Complexity conservation.

Finally, another useful mental picture on why Incompletely Predictable problems such as Riemann hypothesis, Polignac's and Twin prime conjectures are so difficult to solve is that they require complex mathematical arguments belonging to 'Special-Class-of-Mathematical-Problems with Solitary-Proof-Solution' whereas Completely Predictable problems such as proving even number gaps = 2 and odd number gaps = 2 only require simple mathematical arguments based on mathematical calculus or geometrical gradient method belonging to 'General-Class-of-Mathematical-Problems with Multiple-Proof-Solutions'. Thus Completely Predictable problems can easily be "separately" and "independently" solved whereas Incompletely Predictable problems have to be "combined together" and "dependently" solved with difficulty.

Useful Overall Perspective on Incompletely Predictable entities whereby the word "number" [singular noun] or "numbers" [plural noun] could easily be used interchangeably with the word "entity" [singular noun] or "entities" [plural noun]: Mathematics for Incompletely Predictable Problems makes all mathematical arguments valid and complete in my Ting J., April 12, 2019 paper (based on first key step of converting Riemann zeta function into its continuous format version) and my Ting J., April 26, 2019 paper (based on second key step of applying Information-Complexity conservation to Sieve of Eratosthenes). Nontrivial zeros and two types of Gram points calculated using this function plus prime and composite numbers computed using this Sieve are defined as Incompletely Predictable entities. Euler product formula alternatively and exactly represents Riemann zeta function but utilizes product over prime numbers (instead of summation over natural numbers). Hence prime numbers are encoded in this function demonstrating deep connection between them. Direct spin-offs from first step consist of proving Riemann hypothesis and explaining manifested properties of both Gram points, and from second step consist of proving Polignac's and Twin prime conjectures. These mentioned open problems are defined as Incompletely Predictable problems.

Chapter 4 Jelena 'Shining Light'

I have practiced in the specialty field of Anesthesia, Intensive Care, Pain Medicine, Medicinal cannabis, and Addiction Medicine. When I was training in Anesthesia from 2009 to 2013, I was often told that the ability to effectively communicate with my patients is paramount for good patient care. Useful idiom when practicing medicine: Patients 1st, Doctors and Nurses 2nd, Administration and Regulatory Body 3rd.

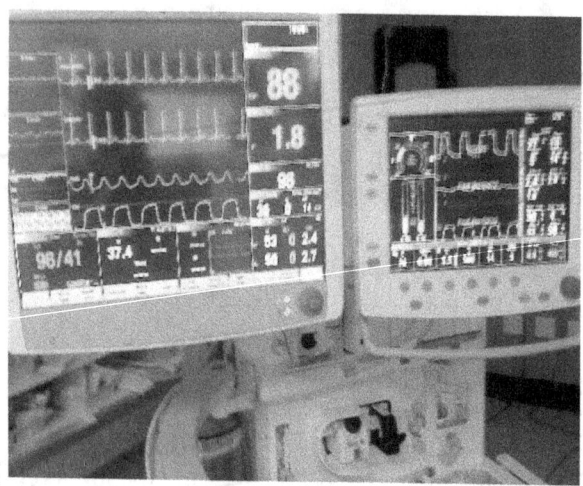

Monitoring.jpg

FIGURE 11. Anesthesia Monitoring.

Rebelling against the pressure-cooker life being a doctor, I have always been privileged to be involved in my spare-time hobby of solving open problems in Number theory of Riemann hypothesis, Polignac's and Twin prime conjectures. I estimate it took me about three years from March 2016 to April 2019 to obtain rigorous proofs for these three open problems. To many, mathematical literature is often seen as an impenetrable wall of logic, symbols and formulas. I recommend this book to readers who want to get a meaningful glimpse of what is behind the wall and how the wall can be penetrated. I have endeavored to write this book describing the artistic, creative and human, and spiritual aspect of mathematical enterprise.

With three older sisters and two younger brothers, I am the eldest son of a domineering father. My mother died from a stroke at 75 years of age in 2016. Being raised on a rigmarole of high expectations and little praise, I developed an unwanted psychological Dependent Personality trait with negative consequences but I resolved this issue in Year 2000 by initiating good father-son relationship with lots of help from family and friends.

On Monday May 14, 2012 my youngest daughter Jelena (meaning 'Shinning Light' in Russian) was born 13 weeks premature with a tiny birth weight of 1010 grams (2.2 pounds). She spent 7 weeks in Neonatal Intensive Care Unit.

From Olaf Helmer and Nicholas Rescher: On the Epistemiology of the Inexact Sciences, P-1513 (October 13, 1959): "In medicine, exact explanation of causes of diseases, concise diagnosis and absolute predictability of outcome of treatment are difficult, if not impossible!" Thus there is a popular saying "Medicine is an inexact science". Neonates do feel pain and require analgesic relief. Remifentanil is a potent, short-acting synthetic opioid analgesic drug with an effective

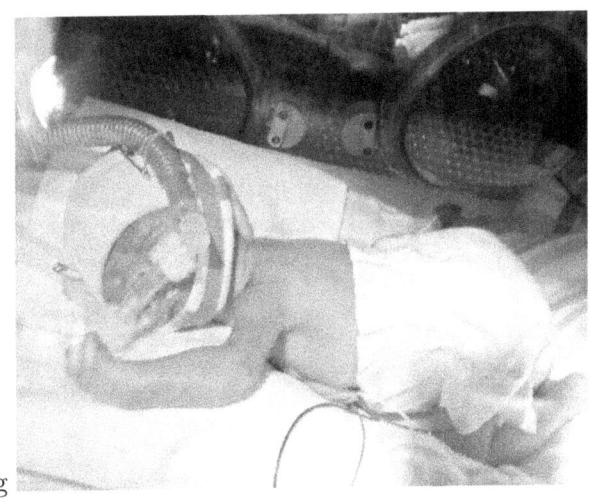

Nasal CPAP Treatment.jpg

FIGURE 12. Jelena Nasal CPAP Treatment.

biological half-life of 3 to 10 minutes. As this drug is esterase metabolized, it is not dependent on immature liver enzymes for metabolism. Therefore its theoretical advantage is that it provide superior analgesia of an opioid without causing prolonged respiratory depression.

For a good cause, my wife Jocelyn and I enrolled Jelena in the premi-remi study on May 21, 2012 at 28.2 weeks gestation during her PICC line insertion procedure in left ankle saphenous vein. With primary aim to determine efficacy of remifentanil infusion for alleviating pain in neonates requiring insertion of central venous lines for their medical care, this study is similar to the study "Remifentanil for percutaneous intravenous central catheter placement in preterm infant: a randomized controlled trial" by (Lago P, 2008). Thus Jelena became part of history contributing data as one of the recruited neonatal subjects in this study based on randomized double-blind controlled clinical trial.

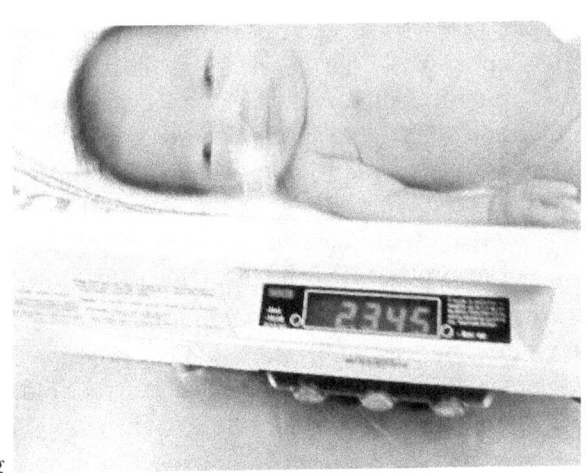

Weight 2345 grams.jpg

FIGURE 13. Jelena Weight 2345 grams

With a feeding tube up her nose, Jelena weight was exactly 2345 grams on July 13, 2012.

Book 2 Three Open Problems by Riemann and Polignac

Chapter 5 Medical Conspiracy and China Economic Miracle

{*Publication Warning: Timing of events in this chapter are altered with real identities of people, places, businesses, entities, and so on hidden or fictionalized so that they cannot be easily recognized or discovered. Any potentially libelous claims and statements that might invade a person's privacy are altered and minimized.*}

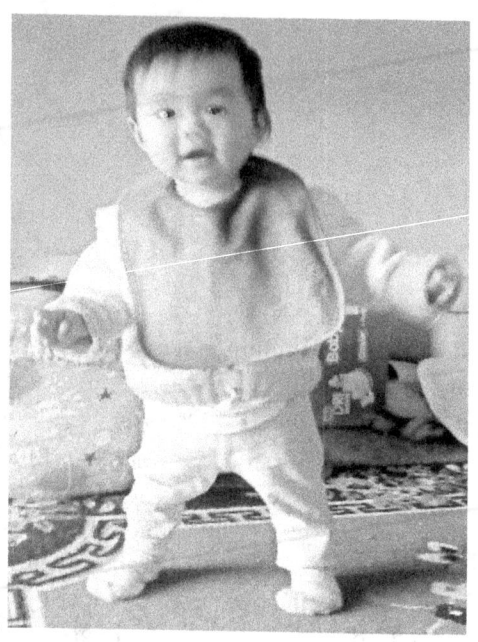

year old Jelena standing.jpg

FIGURE 14. One year old Jelena standing.

Jelena was born 13 weeks early on May 14, 2012. By her first birthday, she is an inspiration and a fighter developing into a normal healthy child. It is a no-brainer I will always be proud of her. I have to stop my full time Anesthesia career in 2013 to help look after my five children. I continue to do some Anesthesia in rural parts of Australia until 2016. At beginning of 2019, my medical career and livelihood were in jeopardy. I can hardly support my family financially as I am unable to work casually in general practice due to regulatory requirements. These mainly arise from unwarranted allegations levelled against me (now using my Pen Name Dr. Bernhard Helpful below) by Big Brother Health Department for practicing Addiction Medicine.

This practice by Dr. Helpful in 'Sleepy Suburb' relates to dutifully combating the rampant and endemic society illness of drug addiction in a chaotic en masse manner for up to 123 difficult and vulnerable patients in 2017 and 2018 when their 101 year-old Dr. Elderly who has been practicing Addiction Medicine in his large Dosing Clinic catering for 175 Drug Dependent Persons and for 20 years was graciously retired off on Christmas Day December 25, 2017. Dr. Helpful was also instrumental in providing common-sense and fair opinions to the assessing psychiatrist in late 2017 on Dr. Elderly's fitness to practice medicine such that Dr. Helpful advocate he should continue to carry so-called S4 and S8 license endorsements for dangerous drug but only use these drugs in the general population [and not for Addiction Medicine]. Here is the Medical Conspiracy. It was subsequently discovered in second half of 2019 that Big Brother Health Department has

year old Jelena birthday.jpg

FIGURE 15. One year old Jelena birthday.

been seeking a scapegoat to blame for a few cases of drug overdoses which may be caused by Dr. Elderly with this perhaps contributing to the death of one young woman in 2017 – see further explanations below. One feels that the cliché phrase "Only the good die young" may apply here. This phrase actually means that highly-regarded people who are morally-upright, kind and compassionate tend to die at a younger age than most people do. It originated in a proverb by Greek historian Herodotus (c. 484 BC – c. 425 BC) in 445 BC. He wrote, "Whom the gods love dies young."

The Diagnostic and Statistical Manual of Mental Disorders (May 2013), Fifth Edition, often called the DSM-V or DSM 5, is the more recent version of American Psychiatric Association's gold-standard text on names, symptoms, and diagnostic features of every recognized mental illness – including addictions. The DSM 5 criteria for substance use disorders are based on decades of research and clinical knowledge. Substance use disorders span a wide variety of problems arising from substance use, and cover 11 different criteria:

Taking the substance in larger amounts or for longer than you're meant to.
Wanting to cut down or stop using the substance but not managing to.
Spending a lot of time getting, using, or recovering from use of the substance.
Cravings and urges to use the substance.
Not managing to do what you should at work, home, or school because of substance use. Continuing to use, even when it causes problems in relationships.
Giving up important social, occupational, or recreational activities because of substance use.
Using substances again and again, even when it puts you in danger.
Continuing to use, even when you know you have a physical or psychological problem that could have been caused or made worse by the substance.

Needing more of the substance to get the effect you want (tolerance).
Development of withdrawal symptoms, which can be relieved by taking more of the substance.

For Severity of Substance Use Disorders, DSM 5 allows clinicians to specify how severe or how much of a problem the substance use disorder is, depending on how many symptoms are identified. Two or three symptoms indicate a mild substance use disorder, four or five symptoms indicate a moderate substance use disorder, and six or more symptoms indicate a severe substance use disorder. Clinicians can also add "in early remission," "in sustained remission," "on maintenance therapy," and "in a controlled environment". Physician, heal thyself (Greek: , – Iatre, therapeuson seauton), sometimes quoted in the Latin form, Medice, cura te ipsum, is an ancient proverb appearing in the Bible at Luke 4:23. In harsh judgement, this proverb seems to be applicable to Dr. Elderly. But good Clinical Practice in Modern Medicine boils down to always employing 'Evidence-based Practice' which roughly equates to balancing 'Evidence-based Medicine' with 'Doctor Experience' & 'Patient Expectation'.

Despite the well-known pharmacodynamics mediated phenomenon in development of drug tolerance with prolonged use of certain drugs that belong to the class of Benzodiazepines and Hypnotics, Dr. Elderly's misguided usage of multiple types of these drugs and in large doses for incorrectly perceived 'opioid-sparing' role did perhaps result in a number of drug overdose cases with harmful consequences. Big Brother Health Department was keen to prosecute Dr. Elderly with resulting loss of dignity. Dr. Helpful non-judgmentally rationalize with assessing psychiatrist that Dr. Elderly's unusual practice was caused by poor understanding rather than medical incompetence endangering patient life. Through negotiation, Dr. Helpful was then asked by Big Brother Health Department to initially "cover" and subsequently "take over" the care of Dr. Elderly's patients.

With Opioid Replacement license obtained on May 3, 2017 and extensive pharmacology knowledge gained from qualifications and experiences in Anesthesia, Intensive Care and Pain Medicine from 2009 to 2013; Dr. Helpful commence seeing this large group of patients under "quota of 100 patients" maiden approval for July 1, 2017 to June 30, 2018 period very soon after starting off with the [usual] initial "quota of 5 patients" approval which was unlawfully dated April Fool's Day on April 1, 2017 [whereby this date is before the May 3, 2017 date of license]. This large "quota of 100 patients" was furthermore unlawfully back-dated on October 13, 2017 to that effect by Big Brother Health Department. Note that the overall goal of good patient care is always for Big Brother Health Department to work closely with all doctors in order to prevent & minimize patient harm.

On February 5, 2019 Mr. Auditor from Big Brother Health Department falsify about sending Dr. Helpful a crucial email containing an important attachment; and furthermore on a separate February 9, 2019 email made a false admission that the alleged February 5, 2019 email was actually sent to Dr. Helpful. This despicable act was carried out by Mr. Auditor on February 5, 2019 with intentional, unlawful & impersonating use of [incorrect] email address Dr.Bernhard.Unhelpful@gmail.com instead of [correct] email address Dr.Bernhard.Helpful@gmail.com. In addition, the senior Dr. Boss from Big Brother Health Department made a number of false allegations against Dr. Helpful about some of his confidential patient care not adhering to required regulations. All above primitive blame-game unfair plays have since been elegantly proven by Dr. Helpful. These acts breached the desperately needed shared responsibilities & mutual trust between a regulatory body & doctors when working as a team to effect "harm minimization" or optimize Harm/Benefit Ratio" for patient care. Importantly, these nasty issues were eventually resolved successfully in a mutually acceptable manner. A popular saying goes like this "Time heals all things" will undoubtedly apply to healing of the strained relationship "wounds" that have since existed between Dr. Helpful and Big Brother Health Department.

Although Dr. Helpful has always provide safe treatments for these patients, he did fail to complete some mandatory paper-works due to factors such as time constraint and health practice location change on December 31, 2017 resulting in lack of access to previous medical records for these patients. Being a Christian, Dr. Helpful was grateful to tackle all above mentioned trials and tribulations with faith in God and support from his close friends and family.

Finally, here is Dr. Helpful choice for top three Mega-Achievements by mankind: American astronaut and aeronautical engineer Neil Armstrong (August 5, 1930 – August 25, 2012) was the first person to walk on the Moon on July 21, 1969 and spoke the now-famous words, "That's one small step for [a] man, one giant leap for mankind." NASA Voyager 1 and Voyager 2 space probes both launched in 1977 to study the outer planets with computing power available in these probes less than that in a modern mobile phone.

China Economic Miracle. People's Republic of China as economic superpower with market economy currently calculated in 2019 as the world's second largest economy by nominal gross domestic product (GDP) and the world's largest economy by purchasing power parity. Until 2015, China was the world's fastest-growing major economy with growth rates averaging 6% over 30 years. She is the world's largest manufacturing economy and trading nation; and also the world's fastest-growing consumer market and second-largest importer of goods. She is a net importer of services products.

In 1998 and 2003, Dr. Helpful extensively tour China, Hong Kong and Macau. Building of numerous skyscrapers in all major Chinese cities occur rapidly overnight enabling the largest recorded-in-history human mass migration [from rural China to urban China]. Since the August 8 - 24, 2008 Summer Olympic Games in Beijing, transport development such as on China high speed train running from 200km/hr to 350km/hr with reasonable priced tickets occur at a phenomenal pace.

Beijing, being administrative capital of China and rich in ancient Chinese history, is often symbolized by the Forbidden City (Chinese: ; pinyin: Gùgōng) and the Summer Palace (simplified Chinese: traditional Chinese: ; pinyin: Yíhéyuán). Shanghai is the symbol of China economic superpower success. Shenzhen, the major city in Guangdong Province, is China's Silicon Valley. It was one of the fastest-growing cities in the world in the 1990's and 2000's contributing to China Economic Miracle. The equivalent Silicon Valley in southern San Francisco Bay Area of California, United States of America, is home to many start-up and global technology companies such as Apple, Facebook and Google.

Creed Odyssey in Mathematics and Medicine series

About the Author: Professor John Ting

family 2016.jpg

FIGURE 16. Photo taken in 2016 of author together with his wife Jocelyn, and children Jonah, Joelle, Jethro, Jonty and Jelena.

John Ting is a Researcher on Fundamental Laws of Nature. His novel Hybrid integer sequence A228186 was published in The On-Line Encyclopedia of Integer Sequences in 2013. From 2016 to 2019, he carries out extensive mathematical research with published papers in Number theory on Riemann Hypothesis, Polignac's and Twin prime conjectures. He lives in Australia with his wife and five children. He possesses Medical degree, General Practice qualification, Primary Anesthesia Fellowship Examination and Opioid Replacement license. His work experiences involve the specialty area of Anesthesia, Intensive Care, Pain Medicine, Medicinal Cannabis and Addiction Medicine. His medical publication in 2012 as primary author with the Professor of Nephrology as secondary author include "Supramaximal elevation in B-type natriuretic peptide and its N-terminal fragment levels in anephric patients with heart failure: a case series".

Book 2 Three Open Problems by Riemann and Polignac   Email: jycting@hotmail.com
Dr. John Yuk Ching Ting, 729 Albany Creek Road, Albany Creek, Queensland 4035, Australia